PENGUIN BOOKS

DRAGON SEA

Frank Pope has worked on underwater expeditions all over the world under the auspices of Oxford MARE (Maritime Archaeological Research and Excavation Unit), including the salvage of Admiral Lord Nelson's flagship HMS *Agamemnon*. He divides his time between London and Nairobi.

DRAGON SEA

SEA

A HISTORICAL MYSTERY.

BURIED TREASURE. AN ADVENTURE

BENEATH THE WAVES

Frank Pope

PENGUIN BOOKS

PENGUIN BOOKS

Penguin ...
Penguin Group (...

Penguin Irela...
Penguin Gr... (Australia) ...

Penguin Books I...
Penguin ...

Penguin Books (Sou...

Penguin ...

First published in the United States of America by Harcourt Inc. 2007
First published in Great Britain by Michael Joseph 2007
Published in Penguin Books 2008

1

Copyright © Frank Pope, 2007
All rights reserved

The moral right of the author has been asserted

Set in RequiemText
Printed in England by Clays Ltd, St Ives plc

ISBN: 978-0-141-02921-4

www.greenpenguin.co.uk

Penguin Books is committed to a sustainable future
for our business, our readers and our planet.
The book in your hands is made from paper
certified by the Forest Stewardship Council.

To my mother and father,
for their gift of freedom.

CONTENTS

FOREWORD

Just as the moon lures the tides, the ocean tugs at a man's mind. Her waters can be many things at once: A healer and a killer, a canvas for contemplation, and an unforgiving workplace. For me, as for so many other young men through the ages, they offered an escape. Fresh out of school with a privileged but conventional education, I wanted to avoid the drudgery that seemed to loom ahead, and to go where there were no trodden paths.

The ocean covers 70 percent of the earth's surface, yet even in its most explored region, California's Monterey Bay, human eyes have seen barely 1 percent of its floor. Given such an unmapped frontier, there were many directions I could have taken and many disciplines to which I could have devoted myself. Geologists follow the undersea forces that drive our drifting continents. Meteorologists watch the sea lead the world's weather in a dance of magical complexity. An imperceptible rise in its temperature sends wind whirling into a devastating hurricane, while an unseen shift of a remote ocean current plunges a continent into winter. Biochemists find life in the depths that survives without light and organisms built without carbon. Marine zoologists explore an environment more exotic than any rain forest: Bring up an animal from below ten thousand feet (still shallower than the ocean's average depth), and the odds are even that you'll have brought up a completely unknown species.

By chance I was drawn into studying maritime archeology, which seeks to understand the story of mankind's four-thousand-year-old affair with the sea. Chance, that is, and the peculiarly powerful allure of shipwrecks. Perhaps the contrast of encountering something manmade in so inhuman an element put my senses on alert, but when I saw my first shipwreck, lying at the bottom of a Greek island harbor, it sparked a fascination that wouldn't die. That sunken ship carried a message from the past that had been sealed on the day of its sinking, the result of a storm, a battle, or a tragic mistake—I didn't know which. It conjured up an era of exploration, trade, and adventure now long passed.

Mensun Bound, director of Oxford University's Maritime Archaeological Research and Excavation unit (MARE), was leading the survey of the medieval wreck, which had sunk off Zakynthos Island. I was an eighteen-year-old volunteer. My father, a classical scholar with a special interest in the history of decipherment, had helped sponsor Mensun's first excavation and so, a few years down the line, Mensun agreed to take me on as a dogsbody. He sensed my enthusiasm and fanned the flame. Over the years that followed he became both a mentor and a friend as we worked together on shipwrecks in Uruguay, Italy, Greece, Mozambique, and the Cape Verde Islands.

I began my journey in archeology expecting to find a clear division between good and bad, right and wrong; instead, I found that the line shifted like bars of barometric pressure on a weather chart. This book tells the story of my last expedition with Mensun, which took place in the South China Sea. The trip was different from all the others on which I had gone, for on it I learned that beyond the power to terrify, enrich, and make judgment, the ocean could also lay bare the very nature of man.

South China Sea,
Mid-fifteenth Century

Every day they waited the voyage ahead became more dangerous. South China Sea's ty-
phoon season was approaching and the captain was growing increasingly anxious, but the
owner refused to leave Van Dong before the ship's holds were full of pottery. Only when
the crew saw the land receding behind them did they at last begin to relax, even though the
hull that bore them was riding low in the water.

Five days into the journey the wind died completely. The sailors knew trouble was
coming, warned by both superstition and experience. When the first waves arrived the ship
began a long, lazy roll. With the sea glassy smooth and not one breath of a breeze, the end
of each pendulous swing was marked by the clunk of loose pottery in the holds and the
clatter of cauldrons hanging in the galley. Then the wind began to blow, howling as it
wrestled with the mast and moaning angrily through the stays. At first the crew tried to
harness the gale to make up for lost time, but the storm soon forced them to drop the bam-
boo sails back to the deck.

In the hours that followed the weather only worsened. Then the ship's blunt bow
shuddered into an oncoming wave and, breaching free from the crest, slipped sideways
into a yawning trough. Men on deck who were heaving at the pumps turned and shouted

in terror as a mountain of water loomed over them. With a summit ridged by furious whitewater, the wave scooped the vessel up its near-vertical face and flung it onto its beam ends.

Below in the galley a kitchen girl cowered as a fist of water punched through the window and sluiced over her. She lay dazed on her back, wedged into the valley between wall and floor, waiting.

With agonizing slowness the ship began a slow roll back upright, her heavy holds now burdened with tons of seawater. Then a second rogue wave hit and the ship rolled over again. This time the hull made no attempt to push itself back upright and the sea flooded into the galley, unstoppable and inexhaustible.

Broken masts and sails streamed above the ship as it sank. Soon the girl's struggles stopped. Weightless among the cauldrons and the kettles, haloed by her billowing hair, life faded from her eyes as around her the water grew dark with depth.

When the ship finally hit the seabed it too had stopped struggling. Ropes coiled downward through fleeing bubbles as it eased to rest in the mud, the sails settling onto the deck like the folds of an emperor's gown.

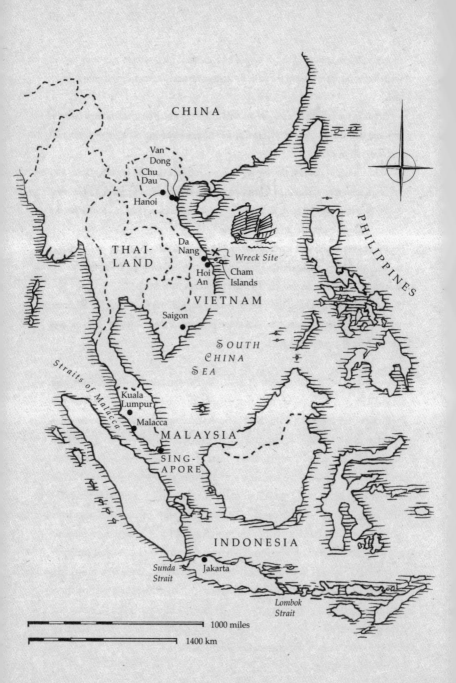

Grandchildren of the Dragon

OFF THE SOUTHEASTERN FLANK of the Asian continent lies the world's largest body of water aside from the five oceans themselves: the typhoon-torn South China Sea. The encircling islands and peninsulas of Malaysia and the Philippines, rather than providing shelter from oceanic winds and currents, serve instead to channel and focus them. Late every summer, low-pressure systems form in the Northwest Pacific and wheel toward the East Asian seaboard, intensifying into tropical depressions, storms, and occasionally typhoons that ravage the coastline. More powerful than any other meteorological force on earth, typhoons are the Pacific Ocean's version of hurricanes, anti-cyclones that can spin at speeds of up to 120 miles per hour. Caused by predictable conditions, typhoons can be avoided by the mariner so long as he stays out of their territory between mid-May and September. Unlike mariners, however, typhoons are not calendar-watchers. They sometimes drop in early, or El Niño can delay their arrival until late June.

Despite the threat of such violent weather, the South China Sea is one of the busiest sea lanes in the world. The Straits of Malacca, the sea's southern entrance that cuts between Sumatra and the Malaysian peninsula, has two hundred ships passing through each day, a figure surpassed only by traffic in the English Channel. The ships—traveling between the East and the West—are forced to use the straits to avoid the hundreds of miles of extra sailing it would take via the Sunda or the Lombok Straits instead. The South China Sea's strategic and economic importance has increased to such an extent that some of the countries bordering it (Vietnam, China, the Philippines, and Taiwan) remain locked in a long-simmering territorial dispute that began after World War II. Much of the military wrangling has centered on the hundreds of atolls, cays, shoals, reefs, and sandbars that are scattered in a thin band running from north to south. These tiny, mostly unpopulated islands—known by the Chinese as the "Tough Heads of the Surging Sea"—sparked eight separate military clashes in the 1990s alone. The two major island clusters, the Paracel and the Spratley Islands, are claimed by both China and Vietnam. In 1974, China seized the Paracel Islands and eighteen soldiers were killed, while in 1988 the Spratley Islands were the scene of a naval clash in which seventy Vietnamese perished. Controlling the thoroughfare is not the only thing that worries the two nations. Some 266 trillion cubic feet of natural gas lie beneath the seabed, as well as an estimated 28 billion barrels of oil.

The resulting political chaos makes the South China Sea an ideal haunt for pirates, who have always thrived in the refuge provided by its convoluted coastal geography. The ancient Chinese navy was harassed by Japanese Wako pirates for three hundred years, until the Chinese struck a deal with the battle-hardened Portuguese to keep the pirates at bay in the mid-sixteenth century. This cleared the way for the rogue Portuguese Franks and a chain of others. Their successors still plague the area. Along with the Gulf of Aden and the Somali coast, the waters of the South China Sea are the most pirate-infested in the world. The International Maritime Bureau's Piracy Reporting Centre, based in nearby

Kuala Lumpur, is trying to combat the problem, but incidents still abound. With its high concentration of shipping, the Straits of Malacca are especially hard hit. Seventy-five attacks were reported in the year 2000 alone.

The legendary brutality of pirates has not diminished with time. In 1998 the freighter *Cheung Son* and its cargo of iron ore disappeared near Hong Kong. Weeks later, fishermen pulled up their nets and found within them the bodies of the vessel's twenty-three-man crew; they had been bound, gagged, and shot. It wasn't until 2005 that Chinese police stumbled on photos taken by the pirates while still aboard the *Cheung Son* as they partied among the dead crew members. In this exceptional instance both the ship was found and the pirates were apprehended. Thirteen men were eventually executed. Hidden by the physical enormity and legal haze of international waters, even huge oil tankers can disappear for good, and pirates are rarely caught. Commercial vessels are not the only targets. Between 1980 and 1985, thousands of refugees fled southern Vietnam on a motley collection of small craft. The "boat people" (as they became known to the world's press) were trying to escape the aftermath of the Vietnam War. The victorious Communist government of North Vietnam was rounding up all those suspected of collaborating with the U.S.-backed regime of the south and putting them into "reeducation" camps, or worse. The refugees were desperate and unprepared for the open sea, and their engines soon broke down or ran out of fuel. As they drifted helplessly, the pirates descended. Hundreds of boat people were kidnapped, killed, or mutilated, and more than two thousand women were raped.

THE VIETNAMESE ARE no strangers to persecution. For more than two thousand years their nation has been plagued by seemingly endless internal strife and foreign invasions. Their most persistent foe has been China, forever a looming presence to the north. In 179 B.C.E., a southern Chinese kingdom called Nan Yueh (Nam Viet) first managed to occupy the country known as Au Lac, in the northern third of what is modern Vietnam. Control of the Red River Delta was the invaders' main objective, for its tributaries laced the region, providing minerals, irrigation, good clay for ceramics, and far-ranging avenues for trade and transport. In III B.C.E., the Han Empire inherited control and began a program of Sinification, systematically replacing the local Vietnamese traditions, nobility, religion, and regulations with Chinese versions. Perhaps most significantly, the Chinese language was imposed for all official and literary expression.

The Vietnamese refused to accept the occupation, and in A.D. 39, two young sisters called Trung led a rebellion and succeeded in establishing an independent state run by the elder sister, Trung Trac. The Chinese quashed the insurrection four years later, but the sisters' heroic stature had become firmly established. For the next thousand years the Chinese struggled for control of the Red River Delta area, exerting influence on their southern neighbors either through administrative rule or cultural hegemony. The Tang dynasty (A.D. 618–906) optimistically rebranded the land Annam, or "Pacified South," and forced the Vietnamese to learn Chinese history and the classic Confucian texts at school. Only those sympathetic to the Chinese Empire were allowed to prosper and a sycophantic upper class developed, but among the farmers and scholars a proud and independent spirit still burned, and they held the legend of the rebel Trung sisters close to their hearts.

Their overt impertinence infuriated the Chinese, whose historians scornfully reported evidence of Vietnamese barbarity. Women, for instance, were given equal status in society, and even daughters could inherit. They worshipped fertility rather than their ancestors. They chewed narcotic betel nuts and their kings wore tattoos. More than anything, however, it was their refusal to accept the Chinese language that irked the occupying powers. For the Vietnamese, their separate spoken language had become both the emblem of their independent identity and a major tool in the resistance.

The rebellion gathered strength, and after a desperate battle against the Chinese a local warrior named Ngo Quyen became Vietnam's first independent ruler in A.D. 938. Warlords still fought among themselves for control, and it wasn't until 960 that Dinh Bo Linh emerged victorious and unified Vietnam. He was succeeded some forty years later by the Ly rulers, who managed to repulse repeated invasion attempts by the armies of the Chinese Song dynasty. When the Tran rulers took over from the Ly in 1225, they did all they could to assert the independence of their own culture. In particular they adapted Chinese characters to form a specifically Vietnamese script, called *nôm,* which continued to be used until modern times.

With the arrival of the Ming dynasty in China came a new desire for expansion of the Chinese Empire, and in 1407 the Ming armies invaded Vietnam with massive force. This was a zero-tolerance campaign. After regaining control, the "army of liberation" set about stamping out all signs of Vietnamese independence. Monuments, books, paintings, temples— all were destroyed as part of a systematic operation. For the next twenty years the Ming emperors ruled over Vietnam with an iron fist, hoping to crush the rebellious southerners forever.

The Ming empire's great desire to expand its territory and influence was not restricted to Vietnam and the south. With nothing to the north or east but frozen plains and empty ocean (save for the islands of Japan), the west became its main objective. However, the land routes westward from the empire were effectively barred by rugged mountain ranges, steep escarpments, and high, barren tablelands. The "silk road" never existed; in reality there was a broken chain of fiefdoms and warlords, each of which exacted crippling taxes from anyone seeking safe passage through their territories. The only way for China to extend its reach was via the South China Sea, and to this end seven great exploration fleets were dispatched across it and into the world beyond.

The fleets inspired awe in all they encountered. The marvel was not simply in the size and number of the ships but what lay in their cargo holds: Aside from bolts of gloriously colored silk, the Chinese merchants who stepped down the gangplanks offered ceramics that seemed impossibly delicate. The material was so thin as to be nearly translucent yet at the same time amazingly strong. As if to deepen their mystery, the pieces would sound a musical tone when struck. Dazzling blue designs contrasted beautifully against a white background of perfect purity. When the first tea sets reached Europe they sparked a collecting craze, and in Africa even isolated fragments of the porcelain were revered by witch doctors as magic totems. Nothing even remotely comparable had ever been seen, and though they were valuable at the time, they have gone on to become ever more so. Blue-and-white porcelain was destined to become the most successful form of ceramic ever created.

WHILE THE VIETNAMESE strenuously resisted what they saw as Chinese impositions such as the Confucian religion or the binding of women's feet, over the centuries they proved willing to adopt lucrative technologies developed on the other side of the border. Inspired by the Chinese, a Vietnamese ceramic industry sprang up—a debt that they acknowledge in their legends. One tells of the Vietnamese artisan Truong Trung Ai meeting a Chinese potter at his kiln in Dau Pho on the Red River in the second century B.C.E. and becoming his pupil. Another, from the early Ly period (1009–1225), recounts the story of three Vietnamese officials learning the secrets of Chinese pottery when forced to seek refuge in Guangdong in southern China during a storm. However, the Vietnamese were unable to match either the scale of production or the purifying techniques that the Chinese had refined, and for centuries the Vietnamese industry was confined to selling to local markets bordering the South China Sea.

As with ceramic production, other elements of Chinese culture gradually diffused into Vietnam's art and literature. Among them was the most powerful symbol in Chinese mythology: the dragon. So deeply rooted is the dragon in Eastern legend, its presence felt in so many Eastern cultures, that its origins are hard to discern. What is obvious, however, is that Chinese dragons are of quite a different breed than the dragons of the West. Rather than lurking under mountains, amassing a hoard of treasure and guarding it with flame-throwing breath, the Eastern dragon is a watery, benevolent beast. Exhaling clouds of steam rather than fire, the mythical Chinese dragon kings, the Hai Lung-Wang, lived in fabulous palaces ten thousand feet below the waves. "In deep waters he is in his most natural place and is at his most powerful," one tale begins. "It would take a brave man indeed to seek him there . . ."

The Vietnamese once referred to the capital of their country as the City of the Rising Dragon, and until 1293 every king had a dragon tattooed on his upper thigh. A popular fourteenth-century myth has the Vietnamese people descending from a union between a dragon and a fairy. The fairy came from the mountains, while the dragon lived in the sea. Opposites attract, and when the two met five thousand years ago their passion was rewarded with one hundred sons. Before long, however, the dragon became restless—he wasn't cut out for the fairy's life in the mountains—and explained to his lover that he had to return to the sea. In an impressively rational divorce settlement, he took fifty of his sons with him. Today, people from all parts of Vietnam refer to themselves as grandchildren of the dragon and the fairy, although a dramatic distinction exists between the highlanders of the fairy's mountains and those who live near the coast.

Ever since the brutal invasion by the Ming armies in 1407, the grandchildren of the dragon have given occupying armies a hard time. The more the Chinese stamped out signs of independence, the more the Vietnamese resisted. An organized rebellion took hold under the leader Le Loi, and after a long and determined fight the Chinese forces were eventually routed in 1428. Le Loi was crowned King Le Thai To, the first emperor of the Le dynasty, and joined the Trung sisters among the Vietnamese rebel heroes still celebrated today.

King Le Thai To and his four successors ruled over a period of peace and prosperity the like of which the Vietnamese had never before experienced. After being brutally suppressed during a twenty-year occupation, the Vietnamese nationalist spirit reemerged from the crumbling remains of Chinese Confucian rule and a free-flowing Taoist love of nature and organic processes grew up. The new Vietnamese king, conscious of the damage done to his nation over one thousand years of Chinese occupation and influence, encouraged the country's artisans, writers, and thinkers to celebrate their new freedom and rewrite many of the founding Vietnamese myths and legends that had been destroyed and purged from communal memory, the story of the dragon and the fairy among them. The result was a great cultural outpouring, producing fabulous new palaces and works of literature, silk paintings, and songs that celebrated the rebirth of a nation. The Le dynasty had launched a golden age. The burst of creativity encouraged industry, too, and the potteries that used the fine clay of the Red River Delta expanded their operations. A small village called Chu Dau, marked by smoldering kilns, their interiors stacked high with blue-and-white glazed pottery, was one of many located throughout the lowlands.

The Vietnamese potteries were aided by more than their country's newfound independence. The great Chinese maritime expeditions of the early fifteenth century had not only successfully spread the word about China's splendor but had also brought back foreign luxuries and, with them, foreign ideas. Fearing an erosion of the values that had made their empire so magnificent, conservative elements within the imperial court began a campaign to persuade their rulers to ban the construction of oceangoing vessels, close the borders, and shut down trade. The enormous cost of the exploration fleets was a powerful argument, and the isolationist campaign succeeded. In 1436, maritime trade was seriously curbed. Then, in 1447, an imperial edict banned production of trade porcelain and prohibited all sales to foreign merchants. Breaking the rules was punishable by death. However, the demand for blue-and-white ceramics overseas was undiminished, and the Vietnamese craftsmen of the Red River Delta began to fill the void, using and developing the advanced techniques of throwing and glazing they had learned from the Chinese. They refused, however, to follow the rigid traditions of shape and style that bound their northern neighbors. Chinese compositions, governed by unbending rules (such as never placing a mythical beast in a landscape), were thrown out the window; Vietnamese inventions were inserted in their place. When they realized that their wares were selling well, the Vietnamese artists gained confidence, and the paintings on the pottery became an impulsive record of the glory of their country's independence and rebellion.

Alas, this golden age did not last. Without a controlling authority against which to resist, two of the most powerful aristocratic families in the Le court, the Trinh and the Nguyen, began to compete for control. After a mere eighty years of independence, the country descended into civil war; the powerful Trinh retained control of the heartland in the Red River Delta, while the Nguyen occupied what is now central Vietnam. The fruits of the nation's creative outburst were destroyed as paintings, books, and documents were burned during the bitter fighting between the two families.

None of the art and literature produced by Vietnam during its eighty years of independence endured; what survived the fires soon disintegrated in the hot, humid climate. Memories of the glorious Le dynasty faded as the civil war festered during the course of the following century. With the arrival of the first missionaries—the advance guard of colonialism—Vietnam's fate was sealed. Toward the end of the nineteenth century the French invaded, followed by the Japanese in World War II, then the French again, and finally—as if to make sure not a trace of the golden age remained—the U.S. Air Force's B-52s carpet-bombed the heart of the north. The only surviving testaments to the period were glazed onto the surface of the ceramics produced in the Red River Delta. But while the highest-quality Chinese porcelains were made for the emperor and collected in his palaces, all of the ceramics produced in Vietnam were loaded onto ships and exported to what today is Indonesia, Malaysia, and the Philippines, to be dispersed among the island villages where they have now disappeared. By the end of the twentieth century, Vietnam's brief flowering had slipped into the realm of legend. There it would have stayed, but for a single unfortunate junk, caught five hundred years ago in the teeth of a typhoon in the Dragon Sea.

Part One

CHAPTER I

The Archeologist

THE FALKLAND ISLANDS were a lonely place in which to grow up, and young Mensun Bound was often left to his own devices. With only a few hundred settlers scattered across Britain's desolate outpost in the South Atlantic, there were not many other children his age. Whenever the bitter winds and slashing sleet allowed, Mensun would walk over the beaches and low-lying hills, all featureless save for sheltering penguins and windblown huddles of sheep, to sit on the westernmost rocks and watch the sea.

Squinting into the horizon he would imagine topsails appearing, followed by mainsails and a dark hull, and fantasize about life on board the square-riggers during the Great Age of Sail, the era of exploration, discovery, and adventure. The slate-gray waves were the perfect backdrop for his daydreams as they rolled in from the storms of Cape Horn, some three hundred miles to the southwest, and heaved themselves onto the rocks. Such storms had delivered hundreds of ships onto the island's shores. Some, like the weather-bleached remains of the *Charles Cooper* that dominated the view from his bedroom window, had been so battered by the Horn that their crews had hauled the leaking hulls up on shore and deserted them. Others had met more dramatic fates and were commemorated by the crosses that scarred the region's maritime charts.

In the evenings by a peat fire, Mensun's father would tell tales of shipwrecks and marooned mariners. Mensun's ancestors had been among the first settlers on the islands, drawn by a desire for a Spartan life, close to the elements and away from people. They hadn't been disappointed. In the words of Robert Fitzroy, the captain of Charles Darwin's ship the *Beagle,* "a region more exposed to storms both in summer and winter it would be difficult to mention." There was no television, no radio; the only contact with the outside world came with the arrival of the supply boat every four or five weeks. Among the luxuries it brought were magazines—*National Geographic* and *History Today*—which Mensun scoured for stories involving the sea.

The South Atlantic permeated every aspect of life on the islands, providing the people with food, work, and contact with the outside world. It also isolated them. As a result, when Mensun was eleven he had to be sent to the mainland to attend school. Relations between the Falklands and their closest mainland neighbor, Argentina, were strained. The South American nation contested Britain's ownership of the islands, so Mensun was sent farther north to the capital of Uruguay, Montevideo. He thrived in the cosmopolitan city, becoming something of a bohemian artist, growing his hair long while nurturing a mounting wanderlust. As soon as he returned to the islands, his school years over, he knew it was time to leave again. Convinced that he was destined for a life at sea, Mensun got himself the only job he could, as the engine-room greaser on a ship, the *RMS Darwin.* His parents tried hard to dissuade him. The *Darwin* was a tramp steamer, her itinerary unpredictable, determined only by the destination of her next consignment, and Mensun would be deep in the hull with a grease gun and oilcan for his entire working shift. But his mind was made up: He wanted to wander free across the oceans and into exotic South American ports, seeking to share the experience of the sailors who'd braved Cape Horn before him.

Mensun's parents need not have worried about losing their son to the engine room. After a year on board, with the vessel moored in the Straits of Magellan, he abandoned ship. Life belowdecks hadn't matched his fantasies of adventure on the high seas. The romantic world of Hornblower was gone, he realized. With only his last paycheck and his duffel bag, he began to hitchhike his way north. Eight months later, in 1971, the Falkland Islander arrived in New York City.

Having left one of the quietest places on earth less than two years earlier, Mensun now found himself in one of the most frenetic. He reveled in the atmosphere of Greenwich Village, where he began to play bass in a band, absorbing as much as he could of the city's energy. The influence of the metropolis would stay with him even decades later in the form of his ever-present jeans, unkempt hair, and unusually determined attitude. But for all that Mensun had adopted New York, a big part of him remained a Falkland Islander. He often felt out of step with the world, as if he had been born in the wrong era. As a result, whenever modern life got to be too much he would retreat into books about the past, immersed in a world that he felt he better understood.

When Mensun decided to go back to school, studying history was a natural choice. His lonely youth and the bookishness it had fostered served him well, and he won a full scholarship to study ancient history at Rutgers University in New Jersey. His aesthetic streak found an outlet too. In the lectures he attended Mensun realized that art, and pottery in particular, offered a window into the past. Hollowed stones, wood, and sewn skins were all used as containers by prehistoric cultures, but woven baskets and ceramics were much more suggestive of the people who had made them. Pottery's durability meant it persisted long after all other artifacts had disintegrated. Fragments of fired bowls dating from as far back as 6500 B.C.E. have been found in Turkey, while figurines and animal models from about 25,000 B.C.E. have been discovered in the Czech

Republic. Except among nomads (for whom pottery was too heavy and fragile to be useful) and those who lived where gourds were plentiful (negating the need for artificial containers), most cultures used pottery in some form. By the time Mensun had progressed from examining the evolution of amphora handle shapes to the painted scenes on Greek glazed pots, he realized he had discovered a passion. He gave up playing bass and took a position as a research assistant at the Metropolitan Museum of Art in New York.

Mensun found he could lose himself in ancient history through studying pottery in a way he never could simply by reading about it. The earliest sophisticated ceramics were made by the Greeks. At first they had depicted figures in black against the red ocher of the clay, but sometime around 530 B.C.E. they began to reverse this, painting the background black and leaving the figures red. This meant that the artist was painting with shadow, not light, allowing the figures—usually naked— to be rendered with lifelike accuracy. Beautifully painted characters played out stories of Achilles's victories or of cavorting satyrs; Mensun delighted in translating and interpreting these scenes. The more he studied the pieces, the richer their legends became to him. Soon he could distinguish the styles of many painters, such as Kleitias, Pamphaeus, or Epictetus, without needing to look at the signatures with which they adorned their work.

Mensun found himself at home in the academic life and soon knew he wanted to contribute to it. Much of ancient art had been discovered on archeological digs prior to being displayed in a museum. By studying archeology rather than art history, Mensun felt he could put himself in the front line of the quest for knowledge, interpreting the past when it was first discovered rather than reinterpreting museum pieces and artifacts from established collections. In 1976, at age twenty-three, Mensun graduated with high honors in ancient history and applied for a master's program at Rutgers that combined classical archeology and art history.

It was obvious to Mensun that there were discoveries to be made. He decided he was going to become an explorer, but devote his life to revealing the uncharted past, not uncharted continents. To this end he continued his studies in classical archeology, setting his sights on Oxford University's famous Institute of Archaeology. In 1979, his master's degree complete, he was admitted to Lincoln College to study under John Boardman. Boardman was a leading classical archeologist and art historian and Mensun was overjoyed to be taken on as one of his protégés. But when Mensun arrived at Oxford and proudly walked into the Institute of Archaeology, he was in for a shock. The receptionist looked up the postgraduate's name on her list and found it in John Boardman's group. "So you're one of his lot," she said. "You're not welcome here, you know." Mensun spun out the door before the woman could tell him she'd been joking. Soon he learned that there was truth in her jest, however. Archeology at Oxford was riddled with rivalries and resentments. At their root were differences of opinion as to what was worth studying or on what to spend limited funds. Was it better to find out more about the world of Homer, Praxiteles, and Archimedes or to chart for the first time the development of stone axes in Scandinavia? The battle lines shifted according to the level at which decisions were being made. Within the humanities faculty the Amerindian anthropologist might be at loggerheads with the Egyptologist but at a higher level, when on a committee having to make a choice between doubling the size of the Cultural Museum or founding a new Institute of Homeopathy the two would undoubtedly find themselves on the same side. One of the many rifts was between classical archeologists—the old guard—and those devoted to the newer fields, such as studying the Celtic cultures of the British Isles. Celtic archeologists, who dominated the Institute of Archaeology at Oxford—worked their sites with forensic accuracy, devoting their attention to the relationships between artifacts as much as to the artifacts themselves. Some regarded classical archeologists, who often worked with objects recovered before the latest techniques had been developed, more as art historians than scientists. The receptionist's rejection was Mensun's introduction to the politics of academia that would haunt his later career.

While still a schoolboy, Mensun had picked up a book called *Archeology under Water,* written by a young American academic named George Bass in 1966, and read it cover to cover many times over. From then on, Mensun had followed the fortunes of Bass, Peter Throckmorton, and Jacques Cousteau, and nurtured an interest in exploring the history of man on the sea. The discipline of maritime archeology was still in its infancy. Many land archeologists were skeptical of the possibility of gaining systematic information from the seabed. It took faith and single-mindedness for pioneers like Bass to develop techniques suitable for underwater archeology. It was also exceedingly expensive. But soon it became apparent that shipwrecks were full of clues to the past. Not only did vessels represent some of the most advanced technology around in the classical world, but the discovery of their cargoes also shed new light on its everyday commerce. Perhaps as important, shipwrecks preserved information in a way that very few land sites could—by freezing a moment in time. On land, historic sites in population centers are often piled on top of one another, while those in deserted areas were picked clean of anything of use or value, then left to crumble over the centuries, slowly sinking as the soil accumulated above them. In both cases, the archeologist is left to determine the period to which each layer of a dig corresponds, and there can be many layers.

The few land sites that were instantaneously entombed provide a much more vivid picture of everyday life. When Mount Vesuvius erupted in A.D. 79, drifts of volcanic dust engulfed the Roman town of Pompeii, completely submerging it. When it was (accidentally) rediscovered in the mid-eighteenth century excavators found the accoutrements of everyday life exactly as they had been left two millennia before, including loaves of bread and fruit preserved in glass jars. They also discovered voids in the solidified ash that, when filled with plaster, created eerie reproductions of cowering bodies.

Shipwrecks take their secrets to the seabed in the same way, encapsulating everything on board at the moment of sinking. Only in exceptional environments, such as the motionless and oxygen-free depths of the Black Sea or the Great Lakes, for example, can ships survive for centuries in a pristine state. Usually the ship's timbers are eaten away by marine organisms and their contents dispersed by waves, currents, and tides. However, as long as they are undisturbed by human hands, the scattered remains lie there, waiting to be deciphered.

FALKLAND ISLANDERS take the sea very seriously and never treat it as a playground. Mensun was no exception—he would never swim or sail in the ocean for recreation. However, inspired by all he'd read about archeology underwater, he enrolled in Oxford University's Sub Aqua Club and began learning how to dive.

Before the advent of scuba (self-contained underwater breathing apparatus), the underwater realm was the preserve of sponge, scallop, and pearl divers. Initially these foragers would simply hold their breath—a technique now known as free-diving. People have been harvesting the seabed in this way since 4500 B.C.E., and the famous pearl-diving women of Amoy in Japan still do, along with the fishermen off Mozambique who hunt big blue-water fish. Holding homemade spear guns and masks, they are dropped off by a sailing dhow in open water, diving as deep as one hundred feet to await passing prey. They string their catches on a tiny float, and hope to escape the attention of roaming bull sharks while waiting for their tender to find them at the end of the day.

When precious cargoes went down in shallow water, the Spanish plate fleets employed Venezuelan pearl divers to recover the treasure. The legendary prowess of these divers extended to reports of fifteen-minute dives in otherwise accurate documents, a physiological feat matched only by seals and other marine mammals. However, even if such feats (considered impossible by today's experts) were achieved, breath-hold divers could not hope to bring up an entire cargo. A ship that came to grief with treasure on board held far greater incentive than sponges or scallops, and technology was soon brought to bear. By 1828, the English brothers John and Charles Deane had fashioned a hard-hat helmet that worked like an upturned glass, covering only the diver's head and fed by a continous stream of air pumped from the surface. The principle behind the invention would persist to present day diving apparatus though many details needed attention. The main snag was that the diver had to remain perfectly upright; if he did not, the air would spill from his helmet and rapidly be replaced by water. Twelve years later, civil engineer George Edwards came up with the solution: a "closed suit" that joined the helmet to the rest of the dress. Surrounded by so much air, a pair of boots weighing thirty-six pounds and two forty-pound weights on his chest and back were necessary to keep the man on the seabed. The idea took off, popularized and developed by the enterprising Frenchman Augustus Siebe.

Late in the year 1900 a Greek sponge-diving boat, equipped with Siebe's hard-hat equipment, sought shelter in a bay off the island of Antikythera in the Mediterranean Sea. The captain sent down a diver to see what the seabed held. Only minutes later the diver resurfaced in a state of panic. On the seabed he'd come across a horde of frozen men, women, and horses, their features eaten away as though by syphilis. The captain scoffed and went down himself, returning triumphantly with the arm of a bronze statue. The discovery became a dramatic pointer to the possibilities of marine archeology.

The boat upon which the sponge-divers had stumbled once belonged to the Roman general Sulla, who captured Athens in 86 B.C.E. and took many of its treasures back with him to Rome. It held the biggest hoard of ancient Greek bronzes yet found. The Greek nation was overjoyed. Not only had they made their first major archeological discovery in their own country, they had recaptured their own heritage, stolen from them in antiquity.

The salvage of the Antikythera wreck is seen as the first step on the road toward the science of maritime archeology, because Professor A. Economou from the University of Athens was brought on board and took great interest in the objects that reached the surface. However, there was no thought of his visiting the seabed himself. Diving was not considered something a scholar should undertake. In Jules Verne's *20,000*

Leagues under the Sea (published in 1875), the fictional Captain Nemo agrees. Breathing air delivered from the surface was undesirable, for "under these conditions the man is not at liberty." The solution, Captain Nemo suggests in a conversation with Professor Aronnax, is to use the "Rouquayrol apparatus." Invented in the real world by Auguste Denayrouze and Benoît Rouquayrol in 1865, the equipment consisted of a reservoir of air worn on the back that would last for a few minutes. The suit had to be supplied with air from the surface, but the invention would at least give the diver a few minutes of independence.

The idea was appealing and continued to evolve, and in 1918 a Japanese inventor named Watanabe Riichi came up with the idea of using compressed air in a tank with the air released into the mouth by biting onto a special mouthpiece and dubbed it "Ohgushi's Peerless Respirator." But it was the Frenchmen Jacques-Yves Cousteau and Emile Gagnan who were the first to make the system workable. Cousteau had been dreaming of a system that would deliver air to a diver automatically as he breathed, without the interference of conscious thought. In 1942 he went to Paris to find an engineer and had the luck to meet Gagnan, an expert on industrial gas equipment. When Cousteau outlined his idea, Gagnan immediately understood—it was very similar to a demand valve that he'd designed to feed cooking gas to automobile engines. The first prototype of the modern scuba demand valve was ready a few weeks later and by October 1952 Cousteau was in a position to write, in *National Geographic* magazine, that "the best way to observe fish is to become a fish. And the best way to become a fish—or a reasonable facsimile thereof—is to don an underwater breathing device called the aqualung. The aqualung frees a man to glide, unhurried and unharmed, fathoms deep beneath the sea. It permits him to skim face down through the water, roll over, or loll on his side, propelled along by flippered feet. . . . In shallow water or in deep, he feels its weight upon him no more than do the fish that flicker shyly past him."

Cousteau had brought about a revolution, and he made the most of it. In his television series *The Silent World* he accentuated the contrast between the old world of hard-hat divers and the new as his silver-suited aquanauts explored the alien environment, even hiding their air tanks in space-style backpacks. By severing the diver's link with the surface, scuba managed to divorce diving from its practical and industrial past. The new technology opened the seabed up for everybody, though most of those doing the diving were adventurous, thrill-seeking types, not bookish academics. Serious-minded archeologists prepared to venture underwater were still few and far between.

In 1953 my father, Maurice Pope, encountered the newly invented aqualung. Dick Mitchell, a friend from his old Cambridge college, Magdalene, had persuaded him to go for a holiday in Cannes and learn how to dive. As soon as they'd got the hang of the equipment they began exploring and engaging in the fledgling new sport of amphora-hunting. They were successful, and encrusted fragments of three ancient pots adorned the dining room bookshelf of our Oxford home all my life. They were part of the furniture, and for years I barely noticed them or thought to ask my father more about them. It turned out, however, that they had sparked the first ever academic underwater archeological project.

Swimming above the ancient relics scattered across the seafloor near Cannes, my father had immediately appreciated the possibilities. Two months afterward while he was at the British School in Athens with John Boardman (another Magdalene man, who would later become Mensun's mentor), he put the idea of using aqualungs to explore shipwrecks to Sinclair Hood, the man about to take up directorship of the British School. Hood was enthusiastic and asked my father if he would be prepared to lead such an expedition, but he was teaching at Cape Town University in South Africa at the time. Organizing it from so far away would be impossible. Hood conceded but persevered with the idea nonetheless, and began pulling together the necessary elements. The owner of a luxury yacht based in Cyprus agreed to lend his vessel as the mother ship, while the UK's *Sunday Times* newspaper put up the money for a Siebe-Gorman compressor to fill their tanks and Richard Garnett—one of the first generation of aqualung enthusiasts—raised a team of divers.

In August of 1954 my father returned from South Africa to join the expedition as it set out for the island of Chios, selected because of the likelihood of finding ancient wrecks in the area. They chose well, for during the weeks that they searched they located the remains of some twenty-seven wrecks. The results of their preliminary investigations (all that the Greek government permitted them to do) were reported by Richard Garnett and John Boardman in *The Annual of the British School at Athens* in 1956. It was the first ever officially recognized venture into underwater archeology, making way for all those who would later turn the field into a science.

MENSUN WROTE NUMEROUS letters to Bass from his Oxford college, hoping for a position on one of his expeditions, but was always turned away. Bass had his cadre of archeologist-divers, and the teams were too small to accommodate extras with little diving experience. Then, in the summer of 1979, when Mensun was working—"scraping dirt," as it is called—on a classical site outside Rome, he got a telegram from Bass, asking if he had valid seaman's papers. Mensun, in one of his letters, had mentioned that he had worked on a ship after leaving school. He replied that he had not only seaman's papers but a seaman's passport. Bass immediately invited him out to join his team.

First, however, Mensun had to find a way to get to Turkey. What once would have been a short journey from southern Italy to Greece and then on to the Turkish coast was no longer straightforward. In July 1974, Turkey had invaded Cyprus, an island populated by an uneasy mix of people of both Turkish and Greek descent. The Turks invaded after a Greek coup overthrew the island's president, and eventually they secured the northern third of the island for Turkey. Diplomatic links between Greece and Turkey were severed, and crossing from one country to the other by boat was not permitted. The alternative overland route between the two countries, through Croatia, would have taken too long and cost too much, so Mensun was forced to consider covert ways of getting through. Eventually Bass instructed him to be in the port on the

Greek island of Cos at a certain time in a certain location. Mensun was there. A man approached him, and once Mensun had handed over most of his money he was bundled into the hold of a fishing boat, handed a pail, and then locked in along with the fetid fishing nets and cockroaches. "The smell was disgusting beyond words but everything in me was pulsing with excitement," he later remembered.

The boat arrived at the Bodrum docks in the middle of the night and hurriedly off-loaded Mensun before slipping away into the darkness. After an anxious few hours, he heard the muezzin's predawn call from the minarets. Thinking that the Arabic wailing was an alarm call and that his intrusion had been detected, Mensun was nervously hiding behind a wall when one of Bass's colleagues appeared and made contact. They left the docks and made their way to join the team, and Mensun learned that until recently the expedition had been operating without a seagoing boat; their efforts had therefore been shore-based. The team, all of whom were both archeologists and capable divers, had been using small inflatables to reach the work site. However, after the 1974 invasion of Cyprus, the United States had imposed an arms embargo on Turkey, prohibiting the Turks from using American-made weapons in their military campaign. In retaliation, the Turkish government had closed all the American air bases in Turkey. Until that point, Bass had enjoyed a close relationship with the U.S. Air Force base at the city of Izmir, located near his camp. When the American servicemen were evicted, they decided to shed some of their supplies and equipment rather than ship them home. Among the booty was a former U.S. Navy supply boat that the air crews had been using for R & R. For Bass the boat was a windfall, giving him the means with which to explore the entire Turkish coastline, but it also came with a legal catch: The craft was registered to carry only twelve, of which one crew member had to be a certified master and two others had to be qualified seamen. Suddenly Bass's team needed men who could do triple duty, as diver, archeologist, and seaman.

Over the next few weeks, Mensun helped the team prepare the boat for active duty. When it was finally fitted out, they went on a trial expe-

dition to Yassi Ada, a small Aegean island off which Bass had been investigating a wreck. As Mensun was preparing to make his first dive with
the team, Bass sat down beside him. To Mensun's delight, the great man
had decided to escort him on his maiden dive as an underwater
archeologist.

Over the following months Mensun and the team scoured the coast,
living off American Air Force in-flight meals and talking to fishermen
about local wrecks, before finally deciding to excavate a Hellenistic-era
vessel that had sunk carrying a cargo of Knidian amphorae. Mensun
knew by now that marine archeology—combining art, ancient history,
and the sea—was his future. He had been inspired by Bass's creed that
an archeologist's life should be dedicated to expanding knowledge of
lost worlds, and that the seabed offered a place where this learning
could be gained.

Since its birth in the 1950s, underwater archeology had become an
established, if specialized, branch of archeology, with its own academic
journals and conferences. The discipline spread from an early focus on
the Mediterranean and the classical realm to include all oceans of the
world and vessels as recent as Napoleonic-era warships and the iron-
clads used during the American Civil War. Unfortunately, the adventurous image of underwater work played against maritime archeologists,
who found themselves under continual pressure from university departments to prove that they were also rigorous scientists.

Archeology on land had gone through similar growing pains in the
nineteenth century. When British aristocrats on the Grand Tour of the
classical sites of Greece and Persia brought back souvenirs, academics realized that these artifacts contained information that could supplement
the existing literature. Slowly the field progressed from simply studying
the artifacts to studying the context within which they had been found.

Given the difficulties of working underwater, keeping tight control
on marine archeological sites was challenging to say the least, but rapidly began to pay off as high standards of precision were established and
techniques adapted and evolved. The pressure for the field to prove

its rigor remained, however, and university archeologists became increasingly fundamentalist in an effort to distance themselves from the impulsive treasure-hunters who also sought out shipwrecks.

For the underwater realm was being explored not just by archeologists, and the seabed had become a place where knowledge was being lost as well as being found. The spirit of exploration that infused the early "frogmen" was captured in the name of one of their associations, the Club Alpin Sous-Marin. The first generation of scuba explorers all wanted souvenirs from their adventures, and when they came across shipwreck sites, they helped themselves. For years no one realized that this was a problem. The few interested academics were outnumbered both by the souvenir-hunters and by skeptical colleagues. By the time the archeological value of shipwrecks was accepted at the end of the 1960s, nearly all the ancient vessels that lay within diveable depths in the Mediterranean had been ransacked.

Bass and other pioneers lobbied strongly for protection of these archeological treasures, and gradually the souvenir-hunting antics of divers tailed off, though a hard core remained who considered anything on the seabed to be fair game. Some Mediterranean countries, including Turkey and Greece, took steps to protect the shipwreck sites on their coastlines, and others, like Croatia, banned diving activity altogether.

Mensun was hungry for experience and looked forward to rejoining Bass's team the following year. When he discovered that Bass had decided not to go into the field that season his disappointment soon turned into stubborn determination: He realized he was going to have to win his spurs on his own. He joined the expedition excavating the *Mary Rose,* a flagship of King Henry VIII's that had sunk in the English Channel in 1545. After that he enrolled on a project investigating the Roman *Madrague de Giens* (which sank sometime between 75 and 60 B.C.E.), one of the largest ancient wrecks ever discovered. Mensun was good at the work; he understood ships, and he was passionate about their history. But he also made an important realization. While crew members might labor at gathering the data, it was the excavation director who pulled it together and did the interpretation. History books recorded the names of only the generals of great battles and the leaders of great expeditions; with archeology, the director was the one who made the real findings. Mensun was determined one day to head up his own excavations.

ONE DAY in the summer of 1980, Mensun called home to the Falkland Islands and learned that the ship that had dominated the view from his boyhood bedroom window was scheduled to be scrapped. He was deeply shaken by this; the ship was a valuable relic from a crucial era in seafaring history. Built in Black Rock, Connecticut, in 1848, the *Charles Cooper* had helped transform the new world. She was a packet ship, a breed of vessel defined by punctuality. Until that time, ships had cast off only when their holds were full, inflicting long waits on the portside for potential passengers and cargoes. By guaranteeing departure on set dates, packet ships delivered not only emigrants but fresh produce and mail with a reliability that the seas had never before allowed. The *Charles Cooper* had been abandoned when steamships proved even more reliable, and it had lain listing in the shallows of the waterfront ever since. Mensun had always assumed that she was destined for resurrection in a nautical exhibition, not the scrap yard. He decided to launch a campaign to save her. He went to visit Alexander McKee, the English diver who had obsessively searched for the *Mary Rose* and then instigated her recovery. If anyone knew how to save a doomed ship, it was McKee.

Sitting in McKee's library, Mensun was explaining the importance of the *Charles Cooper* when suddenly something caught his eye. High on a bookshelf was a row of six amphora fragments, each covered in marine encrustations. Roman amphorae were a common sight on the Mediterranean seabed, but Mensun saw that one of the fragments, a broken handle, had a distinctive shape. He knew instantly that it was Etruscan, centuries older than the Roman pieces with which it shared a shelf. He paused, distracted. He had never heard of an Etruscan shipwreck, yet the piece obviously had spent a great deal of time in the sea. Mensun asked his host where it had come from. McKee frowned. He had found the piece diving off Italy years before. He couldn't remember the exact place, but he had been taken there by a British diver named Reg Vallintine. Mensun forgot all about the *Charles Cooper*. The packet ship paled in importance beside the possibility of an Etruscan wreck.

The Etruscans were a mystery. Said to have originated in Asia Minor, they lived in an area that is still called Tuscany in their memory. What language they spoke remains unknown, though they wrote in a script that was borrowed from the Greeks. They decorated their tombs richly and furnished them with high-quality painted Greek pottery, and until about 400 B.C.E. they gave Rome its kings. Many other features of Roman life, including the gladiatorial games and much Roman religion, came from Etruria, but as Rome grew the Etruscans diminished, until almost everything about them had been forgotten. Next to nothing was known of the ships that had built and sustained their civilization, nor about any ship that old. If he could track the site down and if it did indeed turn out to be an Etruscan wreck, Mensun knew that it would launch his career.

Reg Vallintine, the man who had taken McKee to the site, was easy to find. He was famous in diving circles, having run one of the earliest dive schools in the Mediterranean and subsequently directed the British Sub-Aqua Club for eleven years. He remembered the site well, for he had realized that this particular wreck was different from the others to which he took his souvenir-hunting clients. He had shown some pieces to the local museum, which they had accepted but never followed up on, and in the end Vallintine simply decided to stay away. Mensun examined the entries in Vallintine's meticulous logbook that recorded the dives he had made with McKee and his group. Small sketches in the log indicated everything that been taken from the seabed, along with the name of the diver who had made the find. When Vallintine showed him some photos from the trip, Mensun became certain that the wreck was what he suspected.

Something more extensive than three faded photographs and some sketches in a logbook would be needed to raise the necessary funds, however. Rather than immediately launching an expensive expedition to try to relocate the wreck, he decided to check that the artifacts he had seen in Vallintine's photos were what he hoped. Tracking them down would be no easy task, for twenty years had passed since the dive, and the souvenir-collecting divers had dispersed across the world. Fortunately, Mensun had made another discovery in Oxford before noticing the Etruscan amphora fragment.

Joanna Yellowlees was studying modern languages at the university. She had decided to learn to dive, and turned up in the same theory class as Mensun. While Mensun was not alone in wanting Joanna's attention, she saw something different in him. She later recalled, "The other divers were loud and vociferous but Mensun just sat thoughtfully at the back sucking Polo-mints, wearing a heavy sheepskin coat and quietly surveying the scene . . . He was different from the other students. He walked to a different beat. He seemed older and more exotic." It wasn't long after they fell in love that Mensun told Joanna the story of the Etruscan amphora fragment, and about McKee's logbooks and photos. Joanna readily agreed to help him in his quest, and for weeks they spent every evening searching through European telephone directories for names that matched those mentioned in Vallintine's logs.

Described in the logbook and accompanied by a thumbnail sketch was a complete warrior's helmet, recovered from the site by a German known only as "Hans." After a number of false leads, the Hans in question was eventually

located and at length was persuaded to let the helmet be examined and
photographed. Very few helmets of its kind had been seen before.
Beaten from a single sheet of bronze, the helmet was designed to cover
the whole head of its wearer and was incised with powerful motifs.
Snakes writhed up from the nose and across the eyebrows, while wild
boars charged down the cheek-shields. Mensun pleaded with Hans to
lend the helmet to a museum in Italy for public display, but he refused.
It was his pension, he explained with a shrug and returned the artifact
to his bank vault.

One of Vallintine's fading photos showed a woman holding up a pot
with an inward-curling rim. Though less dramatic than the helmet, it
was a crucial piece, for with careful examination it could be closely
dated. The log linked the vase to an English diver whom Mensun and
Joanna eventually tracked down in Monte Carlo. She had been through
two name changes, but "Bogey Kane" had kept the vase. It had a distinctive yellow-ocher glaze with a checkered band around its shoulder. With
the vase in his hand, Mensun's suspicions became certainty. It was a
Corinthian *kothon,* dating to about 600 B.C.E.

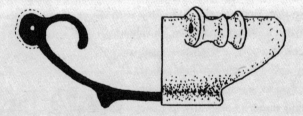

Mensun now had enough hard evidence that the wreck was as old as
he had thought. With funding from the university, the World Ship
Trust, and private donations (including one from my father), he
founded Oxford University MARE and used the symbol of the exquisitely crafted helmet locked in the German bank vault as its emblem.

The wreck lay in a large bay on the northwest side of Giglio Island,
off the west coast of Italy. Divers were ferried to the site by inflatable

boat, and afterward would haul their equipment from the jetty up the hillside to the two ruined villas where they were housed. Compressors thudded and hissed in the front yard, while volunteers toiled away within the crumbling walls. In one room artists made meticulous drawings of barnacle-encrusted objects, while next door conservators treated artifacts in chemical baths. Between shifts, divers sat propped against walls and in empty window frames, logging each dive and every find. In the roofless hall, Mensun held council with excavation elders around maps of the seabed. Though at twenty-eight years old he was younger than many of his team, he commanded respect and kept the volunteers working efficiently. He would later acknowledge experience gained on prior excavations: The *Mary Rose* had taught him about organization, the French about archeology, and George Bass about people. Like Bass, he would readily delegate responsibility to others if he thought they could handle it, especially the endless logistics that accompanied working on a deep underwater site. In other aspects of expedition life Mensun differed from Bass. He decided, for instance, that fraternizing with the team during their downtime was not a good idea, so he ate his meals alone. Fortunately, his lover Joanna Yellowlees was the perfect foil to his detachment, encouraging and organizing the teams with the same grace with which she carried Mensun's dive gear down to the boats, attracting wolf whistles and winks from every male she passed.

The wreck turned out to be everything Mensun had hoped for and more. Greek trade routes were reevaluated in the light of the evidence from the wreck that olive trees had been cultivated in Italy centuries earlier than originally thought. It was discovered that the hull had not only been sewn together, but joined with techniques assumed not to have been invented until centuries later. Hundreds of arrowheads were found scattered throughout the site, and their varying sizes suggested they had been kept for defense rather than trade; piracy in the Mediterranean was apparently already a problem. The ship also showed that Etruscan culture was about more than trade and warfare, for a range of flutes attested to a love of music.

The excavation drew a lot of attention from the world's media. Journalists and television crews often crowded the quay, filming the teams returning from the site and interviewing the divers, who presented their discoveries to the cameras. Enthusiastic crowds cooed and pointed at the ancient artifacts; from the dark interiors of waterside bars, other eyes watched with less innocent intentions.

One day Joanna was working close to the keel in the center of the site. After the vessel had sunk, her upper hull gradually collapsed and the cargo had spilled across the seabed. Storage jars containing thick, black pitch had been on board; it was often used in antiquity as a sealant to protect the ship's timbers. Over the centuries the tar had oozed from the upended containers and entombed several small vases. Excavating such a group of artifacts takes considerable time and skill. At every stage they have to be measured, plotted, drawn, and photographed. At 180 feet below the surface, the site was at the limit of air-diving depth, and each dive could last only six minutes. It took many dives before Joanna was finally able to lift the first pot from the pitch.

When Mensun saw the vase, he recognized the intricacies of the painter's style immediately. His mind reeled. It was as though he had found a long-lost family heirloom. The mannerisms were as identifiable as handwriting. Though the Etruscan artists did not sign their work and their names had been lost to history, specialists had christened them for easy reference. The artist who had created this vase was called "the Little Warrior Painter." Finding his work had already exceeded Mensun's hopes; then Joanna reminded him that there were two other pots still in the tar on the seabed.

The following day, Mensun dived with Joanna to see the other pieces where they lay. When they reached the seabed, the scene before them was almost unrecognizable. The grids were at odd angles, broken swim lines drifted free, and craters scarred the sand. At first Mensun couldn't understand what he was seeing. At high pressures, nitrogen drugs the brain—a syndrome known to divers as nitrogen narcosis. Familiar places seem strange and judgment is impaired. As a result, it took a whole crucial minute before Mensun realized what had happened. Looters had struck during the night. In the area where Joanna had been working, the two half-excavated pots were now gone. Breathing heavily, they looked closer and realized that the looters had cut deeper into the pitch. In what had been a virgin area, Mensun and Joanna could now make out the impressions of five more pots. Perfect, three-dimensional negatives of their shape remained, which meant that casts could be made of their shapes, but this served only to worsen the pain. Mensun would never know the paintings they had carried. On another part of the site, a krater, a large vessel used for mixing wine with water, had also disappeared.

Numb with shock, Mensun and Joanna began to make their way back up the shot line that ran up the rocky slope. In among the weeds they began to see shards of broken pottery. From their shape and color, Mensun realized that they were the remains of the krater, which had been dropped and smashed as the looters were leaving. They were about to investigate further when Mensun checked the contents of his tank. It was almost zero. Joanna's was the same.

At 180 feet underwater, air is compressed to less than one-fifth of what it would be on the surface, but a diver's breath consumes the same volume regardless. A bottle of air that would last an hour on the surface lasts only twelve minutes at that depth. After months of diving on the site, both Mensun and Joanna had become finely tuned as to how long they could spend on the bottom. But in their shock at discovering the devastation, they had taken some deep breaths and depleted their gas supply by a critical amount. With 180 feet of water above them, they could not shoot straight for the surface. They had to remain in the water forty minutes to let the nitrogen escape their tissues without forming lethal bubbles in their blood.

Nitrogen from the air they'd been breathing on the seabed had permeated their body tissues, forced in by the pressure. As they began to surface and the pressure around them dropped, the gases began to expand and leave their tissues, diffusing into their blood, through their lungs, and into their exhaled breath. Mensun could feel his breathing becoming harder. His tank was almost empty and he was having to suck hard on his mouthpiece to breathe. He concentrated on ascending slowly. If they came up too fast the gas inside them would form bubbles that could get stuck in their veins, lodging in their brains or their joints. If it didn't kill them, a bad case of the bends would leave their bodies crooked from the pain. A slow decompression was the only prevention. Mensun tried to calm himself. Panic used more oxygen. He could see that Joanna was also struggling to breathe, only a few bubbles trickling from the regulator in her mouth as she exhaled.

The little gas that remained in their tanks expanded as they rose. By eking it out and ascending faster than they should have, they made it to the emergency air supply that dangled from a decompression stage at thirty feet. Knowing they were out of immediate danger, they clutched each other tightly. Looking into Joanna's face, Mensun saw puddles collecting in her mask, tears streaming from bloodshot eyes. They floated together in the pale blue water until the emergency tanks were almost empty, hoping their bodies would compensate for their rapid ascent while their minds came to terms with what they had seen on the ocean floor.

CHAPTER 2

The Businessman

As THE FINAL SEASON of the Giglio excavations drew to a close in 1986, Mensun was showered with accolades. He was named "archeologist of the year" by the Italian authorities, and photos of him covered the British press, which dubbed him an archeological "superstar," the "Indiana Jones of the Deep." He and Joanna were featured as romantic icons in all the Sunday papers. One weekend a picture of the couple filled the cover of the UK's *Observer* magazine, picturing them clasped in each other's arms as they floated together underwater, he in a suit, she in a dress, their blond hair intertwined.

About to become famous for an entirely different type of underwater endeavor, a little-known salvage diver named Mike Hatcher walked into the ballroom of the Hilton hotel in Amsterdam in the spring of that same year. Christie's was about to start the auction of the "Nanking Cargo," the largest porcelain sale they'd ever attempted, and it was Hatcher's team who had recovered it from the seabed. Though Hatcher didn't yet know it, his treasure hunt was about to be one of the most successful in history.

The auction was already highly controversial. The Indonesian government accused Hatcher of stealing the cargo from their territorial waters, while the Dutch Rijksmuseum had already boycotted the sale, claiming that the wreck of the seventeenth-century Dutch East Indiaman had been destroyed with no attention paid to the historical importance of the site. The International Congress of Maritime Museums went further, stating that "the cargo of the *Geldermalsen* has been looted without concern for context and its commercial sale will entirely destroy the wreck." But with media attention at fever pitch and with twenty thousand people having lined up in the rain to view the cargo before the bidding started, there was no way a little controversy was going to stop Hatcher. He sat down in the front row of the audience, eyes gleaming as he smiled across at his partners, fellow treasure-hunter Max de Rham and a small, sharply dressed Malaysian-Chinese man named Ong Soo Hin. The business brains behind the salvage operation, Ong returned Hatcher's smile with a nod.

Mike Hatcher had been orphaned while still a baby and adopted by Barnado's, a British charity established by Dr. Thomas Barnado in the mid-nineteenth century. When Hatcher reached the age of fourteen he was placed with a family in Australia to finish his education. On leaving school he drifted from job to job, working in construction and demolition, before buying himself the boat he had always dreamed of owning. As he motored north from Australia, he began picking scrap metal from the hulls of wrecked ships that he passed along the way. The wrecks were full of old wiring and piping made of copper, all of which fetched a good price. Encouraged, Hatcher began to take on larger projects, blowing open cargo holds with explosives in order to salvage their contents. He was physically strong, determined, and resourceful, all essential qualities in his new trade, but what he lacked was a businessman's head. Although he had already pulled off several large salvages successfully, Hatcher had always seen his partners walk off with the lion's share of the profits.

IF ANYONE had a businessman's head, it was Ong Soo Hin. Born to Chinese parents on the tropical island of Penang, located off the west coast of Malaysia, Ong was educated in a British manner, topped off with two years in London itself. He returned to Malaysia in the early 1970s—when Mensun Bound was strumming bass in Greenwich Village—to work as an accountant for an established firm. Intelligent, organized, discreet, and always immaculately groomed, Ong progressed fast. Before long he was keeping the books for one of the richest men in Malaysia. As the man grew older he passed some of his fortune to his sons, who inherited their father's trusted accountant along with the money. Ong's responsibilities soon grew beyond simply keeping the books, and he began to give the three brothers financial advice, suggesting projects and investments that invariably produced a good return. Armed with cultured English charm and Chinese entrepreneurial zeal, by his late twenties Ong had become adept at spotting and exploring business opportunities that others had overlooked.

In the late 1970s, Ong heard about a salvage company owned by the brother of the Malaysian prince of Selangor. The owners had argued with the salvager and now wanted to sell; until then, business had been going well. Inspecting the company reports, Ong noted with surprise that there were ships lying on the seabed with large, pristine cargoes of rubber, tin, and lead. None of these products corroded in seawater, and all were valuable; tin, for instance, was worth $43,000 a ton in 1978. Ong bought the company from the prince's brother, and then arranged to meet the salvager himself, Mike Hatcher.

Ong's arrival gave Hatcher a fresh start. In an industry notorious for attracting con men and hustlers, the Malaysian businessman offered a straightforward deal. Knowing that taking a small percentage over a longer period would bring him higher returns in the long term, Ong did his best to help Hatcher succeed. He provided the salvager with a boat, as well as with the compressors, explosives, and dive equipment that Hatcher needed. He also made sure the vessel was well supplied with fuel and food. Then Ong set him loose, turning back to his main investments, confident that Hatcher's keen nose for valuable cargoes would ensure that his speculation made money. That was, after all, the only reason for his interest.

For three or four years Hatcher continued to work as he had been doing, scouring the coastlines of Indonesia and Malaysia for wrecks. Then he started running short of good targets. Most of the shallow-water wrecks had already been salvaged, and the deep ones were too difficult or expensive to reach. When the price of tin began to plummet at the beginning of the 1980s (due to a fall in demand coupled with a surge in production from Brazil and China), the wrecks Hatcher had been planning to work became uneconomical. He was forced to cast his net wider.

Until then he hadn't considered salvaging old wrecks, reasoning that even if they contained goods that might still be valuable today, they had sunk so long ago that they would be all but impossible to find. Then he heard about the Dutch East India Company. During the seventeenth and eighteenth centuries the company had made more than eight thousand voyages to and from Southeast Asia, 644 of which came to grief. The company records—including the locations in which the ships had sunk— were kept in Amsterdam, and were available to the public. All Hatcher needed to do was to find out if any had carried cargoes that were both precious and unaffected by centuries of immersion in seawater. He hired a researcher to compile a list of possible targets. One of the ships on the list, the *Risdam,* caught his eye because the sixteenth-century vessel had been carrying more than twelve tons of tin. The price might have fallen, but it was still a valuable cargo, and one he knew he could sell.

Different wrecks call for different ways of searching, most of which involve the captain running his vessel in parallel lines over the search area while towing some sort of detector. A magnetometer is useful for detecting metals at a distance by sensing local distortions in the earth's magnetic field. Modern ships made of steel show up clearly, but old wooden vessels can also be detected, especially if they sank with cannonry, whose iron can act as a beacon even if buried deep under mud or sand. Side-scan sonar uses sound waves to create an image of the seabed from which an experienced operator can spot an unnatural-looking formation. Neither method is infallible, however: A magnetometer can be thrown by magnetic rock and will miss a wooden ship lacking substantial iron fittings (and tin, being nonferrous and nonmagnetic, would not be detected), while a side-scan is difficult to interpret on coral reefs where strange shapes are the norm. In clear water some wreck-hunters simply tow divers, trusting the human eye to spot anomalies. Whatever the method by which it is found, every target discovered during a search has to be investigated. Some turn out to be modern wrecks or discarded machinery, some simply misleading formations on the seabed, while a few are wooden wrecks—though usually not the ones being hunted.

The *Rísdam* was recorded as having sunk off Pulau Batu Gajah, opposite Mersing on the Malaysian peninsula. While searching for her remains, Hatcher detected a possible target. Divers were sent down to investigate, and found the seabed was strewn with old porcelain dishes and bowls. The vessel's hull had long since rotted away, but below the barnacle- and coral-encrusted upper layer of pottery were some pieces that still looked as good as new. Though Hatcher knew nothing about ceramics, his salvager's instinct told him to bring some up to be valued, just in case. He was in for a pleasant surprise: When he took a dish to Christie's auction house in London, the assessor told him the dish was not worth the $20 that he guessed, but closer to $2,000. Hatcher headed straight back to the reef where he had found the dish to collect the rest.

When Christie's sold the "Hatcher Collection" in a series of auctions throughout 1984, it was their turn to be surprised at the value of the porcelain. A vase covered in barnacles that they had valued at $100 went for $1,400. The combination of Hatcher's rough-hewn appearance, a shipwreck, and the delicate beauty of the porcelain had fired the public's imagination. Hatcher alone made over $3 million. Ong was delighted above and beyond the money he had made—Chinese porcelain was far more aesthetically pleasing than scrap metal, and the publicity and drama of the sale had been exhilarating.

What happened next is enshrined in treasure-hunting lore. The archives of the Dutch East India Company were found to contain the records of a ship called the *Geldermalsen,* which sank early in 1752 when her captain made a navigational error and struck a reef. She had been en route to Holland with a cargo of porcelain. The *Geldermalsen* was far from unknown in scholarly circles. It had already caught the attention of a Dutch maritime archeologist named Christian Jörg, who had studied the vessel for his doctoral dissertation. Not only was the porcelain from a period of the Qing dynasty (1644–1911), which was interesting from an art-historical perspective, but the investigation into the sinking, which happened in good weather and in broad daylight, opened a window onto the inner workings of the Dutch East India Company. A detailed archeological excavation of the wreck would have completed the picture, but no academic had yet managed to go out and find her remains. Hatcher needed no encouragement. He teamed up with fellow treasure-hunter Max de Rham and went to track it down. Lured by the prospect of profits and without having to convince university funding bodies, they got there before the archeologists. In 1986, Christie's of Amsterdam took on the consignment, giving it the title "The Nanking Cargo," choosing to name the sale after the type of pottery rather than the ship.

Aside from the legal problems, Christie's was worried about flooding the market. The Hatcher Collection had convinced the auction house that there was interest in shipwreck porcelain, but those particular pieces had been sold in four separate sales over the course of a year. A consignment of 160,000 pieces to be auctioned at one event was an entirely different proposition. However, Christie's had one advantage now that it had not had with the Hatcher Collection: a story. Supported by Hatcher, Christian Jörg turned his thesis into a book on the *Geldermalsen.* The tragic circumstances of the captain's fatal navigational error were revealed, together with the intriguing tale of a missing chest of gold that the most senior survivor, a boatswain, had been accused of stealing as the ship went down. The man had escaped execution or imprisonment, but the charges of theft had plagued him until his death.

Christie's marketed the sale as "the cargo that arrived at the Amsterdam dockyards 233 years late." The original manifest was published, and on the shelves of the viewing hall buyers could view items that had last been seen leaving Canton on December 20, 1751. Sealing the sale's romantic image as treasure from the deep, gold ingots sparkled on the cover of the auction catalog. In the final days of Hatcher's salvage operation, a diver had probed the sand outside the hull and hit something solid. It was the chest of gold, apparently dropped overboard by somebody while trying to make good his escape from the sinking ship.

Christie's estimated that the sale would net $6 million, but, again the bidding surpassed all expectation. Each piece sold for an average of four to five times the estimated price. The 126 gold ingots were a star attraction. Sitting in the front row with his partners Hatcher and de Rham, Ong eyed the gold eagerly. Though the bullion value of the ingots was only about $4,000 each, when the bidding started Ong bought one—for $82,000, twenty times its nominal worth. He'd caught the shipwreck bug, treasure fever. The Nanking Cargo turned out to be the most successful auction of its kind and the second biggest that Christie's had ever orchestrated, winning them 37 million guilders (approximately $20 million).

The sale's success did not delight everyone. The Indonesian government still maintained that the wreck-site on Admiral Stellingwerf Reef (off Bintan Island, southeast of Singapore) lay within its 120-mile "zone of economic influence." Hatcher had refused to recognize their claim, countering that zones of economic influence did not specify shipwrecks as a defendable resource. The Dutch government also joined the fray. In 1795 the Dutch East India Company had been nationalized after becoming bankrupt, and the government still laid claim to all the company's ships and their remains, but according to the journal *Historical Archaeology* Hatcher appeared to have concealed the identity of the vessel whose cargo he was selling. The only artifacts that carried the company's seal—two cannons and the ship's bell—were added to the auction at the last minute. The government felt that Hatcher had intentionally hidden the ship's provenance until the last moment. In the end they settled out of court, for a 10 percent payoff, while the Indonesian government also eventually dropped its charges.

But as the legal furor surrounding Hatcher calmed, the academic outrage began to flare. In the world of maritime archeology the *Geldermalsen* sale represented a tipping point, pushing scholars into battle against salvagers. Although known to be historically important, the wreck had been ripped from the seabed without regard for the insights it might reveal. The academic archeological press condemned the salvage as pillage and the accusations flew. There was a disparity between the cargo that had left Nanking in 1753 and that which had arrived in Amsterdam in 1986. Not everything listed in the *Geldermalsen*'s manifest had appeared at auction, yet nothing was left on the seabed but broken remains. The missing pieces were all from less desirable categories, such as teacups, 32,500 of which never appeared. Some alleged that Hatcher, scared of flooding the market and not wanting anyone else to do so either, had destroyed the remains of the wreck. Whereas archeologists might once merely have frowned upon salvagers, they now became militantly opposed to sales or projects that could lead to similar destruction. Worse still, the scholars foresaw that the publicity surrounding the auction and its runaway success would inspire more treasure-hunters to loot other sites in the hopes of winning their fortune.

They were right to be worried. The success of the *Geldermalsen* auction had come hard on the heels of another treasure-hunting sensation. The previous year, off Florida, the American treasure-hunter Mel Fisher had discovered the Spanish *almiranta* (vice-flagship) *Nuestra Señora de Atocha*, which sank in 1622. Fisher's lifetime of searching had been rewarded with a haul of $400 million worth of precious gems and metals, the richest such find in history. Together the two shipwrecks made it seem as though the seabed were paved with gold and a rash of new treasure-hunting companies sprang up in their wake. By the early 1990s, twenty-five new outfits were being formed every year, their total financing reaching $100 million. With such resources behind them, a new army of treasure-hunters began scouring likely coastlines and shipping archives, searching for the next big thing. Unfortunately, the only method of discerning a wreck's age, nationality, or identity is to start digging it up, and for every shipwreck on the seabed with treasure in her holds there are hundreds with none. Uncovering a shipwreck exposes it to oxygenated water and decay, breaking the seal formed by a veneer of sand, mud, or coral, and even a quick investigation starts an irreversible process of destruction. As the new wave of treasure-hunters searched, they systematically destroyed the time capsules that archeological investigators had learned to value so highly.

The only solution, reasoned a committee of archeologists and management experts at the convention on "Underwater Cultural Heritage" hosted by the United Nations' Educational, Scientific and Cultural Organization in the early 1990s, was to ban the sale of antiquities that had come from the sea and so cut off the market that was fueling the destruction. A similar ruling had saved the African elephant when the trade in ivory was banned in 1989, and there seemed no reason why it should not also save endangered shipwrecks. To cut the demand that was fueling the destruction, UNESCO deemed that it should be illegal to buy or sell any artifact from a shipwreck over one hundred years old. The treasure-hunters would no longer have any incentive to go to sea, and the world's wrecks would remain untouched until their eventual investigation by archeologists. Each such excavation might take as long as ten years, but at the end the information would emerge complete. Most of the major university archeological departments supported the idea. Any excavation that ended in the sale of artifacts was decreed unethical, and any archeologist involved in such a project was deemed a traitor to the cause, or worse, a treasure-hunter.

The maritime nations of the developing world were seen to be most at risk from predation by treasure-hunters. Not only were a large proportion of the world's wooden wrecks concentrated there (a result of Europe's push to profit from the lands they'd discovered), but poorer nations were also most likely to be tempted by the payoffs offered by salvagers. Though the UNESCO convention on shipwrecks did not immediately win enough signatories to become law, its proponents began to lobby the governments of vulnerable countries, making the case that treasure-hunters like Hatcher were robbing them not only of their history but also of a substantial resource. In the long run, the archeological information that lay in shipwrecks and the revenues from museum displays would outweigh the short-term profit from selling any treasure that was found. They had to protect their heritage, or lose it. Governments worldwide—and especially in Southeast Asia—were persuaded to tighten up their salvage laws.

BY BACKING HATCHER and his *Geldermalsen* cargo, Ong Soo Hin had fully stepped out from the safe confines of accountancy and into the world of treasure hunting. The auction made him an instant millionaire, and his investors were thrilled. Ong began to enjoy his good fortune and took up hobbies suited to his new lifestyle. None of them held his attention for very long. He bought a low-slung Austin-Healey sports car, but once he had mastered it he hardly touched it again. He bought a yacht, then lost interest in sailing. He trained obsessively when introduced to tennis—until he beat the man who had taught him to play, after which he never picked up a racket again.

Though Ong apparently loved the profit margins and the glamour, the legal uncertainties involved in projects like Hatcher's were problematic. Hatcher was often working in disputed waters, and it went against the grain to waste time seeking legal permission. "My motto," he famously declares, "is finders keepers." Ong knew that the financial implications of having one of their vessels or a recovered cargo impounded or confiscated would be severe, and there were occasions when things might even have gone one step worse. In 1992, in the Gulf of Thailand, Hatcher was halfway through salvaging a cargo of porcelain when he was confronted by the Royal Thai Navy. Hatcher argued bitterly that he was outside Thai territorial waters, and for three days was circled by six warships—Ong eventually managed to defuse the situation via satellite phone from Malaysia, persuading Hatcher to yield rather than risk the boat.

From that moment on, Ong's business brain began to mull over more efficient ways of making money from shipwrecks, somehow removing the uncertainty of international uproar. Though laudable, it was a contemplation that would lead slowly but surely to the biggest mistake of his career.

CHAPTER 3

The Catch

TWENTY-TWO MILES off the Vietnamese coast, close to the ancient and picturesque trading town of Hoi An and some five miles northeast of the Cu Lao Cham Islands, Captain Trang ordered his crew to heave the trawl nets overboard. They watched the battered foam buoys drift away in the sampan's wake, and waited until the nets below tugged at the lines before they retreated onto the polished hardwood planks of the stern cabin. Smoke and scented steam wafted up from a hatchway to the hold, where a pot was being stirred on an open fire. Bowls of Cao Lau broth appeared, glasses of rice wine were raised—*Chuc Mung!*—and swallowed in one gulp. As they slurped their soup, Trang glanced up occasionally to check around but left the boat to its course, the puttering of the engine overlaid with laughter and the singsong of conversation.

Half an hour later, all was quiet apart from the beat of the engine. Sprawled in the low cabin, Trang and his crew were sleeping off their lunch. They didn't notice the aging floats on the nets beginning to sink. They slept on even when the ropes that stretched from the sternposts suddenly stiffened and creaked. The captain shifted as the engine's note dropped a tone, but he didn't open his eyes.

When he woke an hour later, Trang realized the floats had disappeared, and he growled angrily. The nets were designed to hang near the surface, not drag along the seabed. They might well be damaged, and the trawl would be useless. Still puffy-eyed from sleep, the crew hauled at the lines, straining and muttering under their breath that they must have caught a monster. When the nets finally broke surface they were filled with mud from the seabed, but the dark ooze was punctuated with strange shapes and patches of white. Trang pulled one of the shapes loose. He dipped it back in the sea to wash it clean, and a rice bowl emerged. When the crew saw the blue-lined bird that was painted on the off-white clay of the interior, they clucked with curiosity and began to pick out other pieces from the net and rinse them off. Soon the deck was spread with bowls, jarlets, small dishes, and a few other forms they didn't recognize.

The catch was some consolation to Trang, for although the pots looked a little old, he thought they could be polished up and sold. The few that were broken or stained he threw overboard. Then he ordered his men back to the work of catching fish. The crew fixed more floats to the nets, laughing at what the waters sometimes chose to offer up.

Trang was disappointed when he took his pots to the village market. None of the vendors were interested in them, though one suggested he take them to an antiques shop in Hoi An. Trang did so and to his surprise the dealer there bought them all, and for a good price. Trang made more money from his find than he would in a week of fishing. He returned immediately to his crew and told them to keep their mouths shut. They would be fishing for pots from now on.

That night, with incense smoke curling in the candlelight, Trang stood at a narrow altar in the corner of his house and while chanting in a thin, high voice offered fruit, flowers, and a glass of rice wine to the Emperor of the Sea, the dragon lord on whose generosity he depended. Between the second and fifteenth centuries, what is now central Vietnam

was inhabited by the Cham people. For the Cham, as for most Eastern cultures, the dragon was the most powerful symbol of them all. The Cham paid homage to a dragon god whom they knew as the Guardian of the Sea. Statues of the dragon depict him clutching a pearl in his fangs. According to Cham legend, the fierce-looking creature was benign only because he dared not open his jaws, for fear of dropping the pearl that was at once his treasure and the source of his power.

Chinese dragons also value their pearls above all else, and are locked in an eternal chase for them across silk paintings, around vases, and down the streets of every major city in the world during Chinese festivals. A legend explaining the origin of their obsession tells of a poor boy named Xiao Sheng who long ago found a beautiful pearl in the grass near his home. Wanting to keep it safe overnight, he hid it in an old rice jar that was empty, save for a few grains. When he opened the jar the next morning to show the pearl to his mother, the jar was full of rice. Astonished, he and his mother emptied the container, then replaced the pearl along with a single grain of rice. The same thing happened. The magic jewel made anything around it multiply. Realizing they would never go hungry again, Xiao Sheng and his mother were overjoyed and shared the good news with their fellow villagers. Unfortunately, one man grew greedy when he heard about the pearl's power and resolved to steal it that very night. Xiao Sheng woke as the thief approached but was caught as he tried to escape. Terrified of losing his treasure, the boy swallowed the pearl. Immediately Xiao Sheng began to glow as a great heat rose inside him, and the thief ran off in fear. Desperate to quench the fire, Xiao Sheng dove into the river and began to drink, but the pearl kept burning inside him. The boy kept drinking and began to grow, gradually assuming the shape of an enormous dragon. The dragon was as kind and generous as the boy he once had been, pouring much-needed rain from his water-filled body onto the villagers' crops. His ferocity was reserved only for the greedy, like the villager who had tried to steal his pearl.

Captain Trang was a practical man, but he had witnessed typhoons that the dragon sent when angry and knew men who had perished in them. He prayed and gave thanks, reaffirming that the pottery that he had found in his nets had been a gift from the sea, not stolen. Early the next morning he burned more incense sticks on the bow of his sampan as he motored out. He watched the islands shift behind him, his eye unconsciously measuring outcrops against areas of color and then flicking back to the mountains of the mainland. Like his father, he could pinpoint patches of water almost instinctively. Though he didn't know exactly where his nets had snagged the pots, he could repeat the course he and his men had followed. This time he sank his nets deliberately.

The crew watched the ropes eagerly all day, their usual siesta forgotten, but time after time the nets pulled up only mud. Trang had to use all of his rope to reach the bottom. Though he could control the track he was making on the surface, he could not predict exactly where his nets would land on the seabed. Unseen currents were snatching at his lines, pulling them in different directions at different depths. He returned empty-handed that first day, and again on the second. But on the third his perseverance was rewarded.

After a week or two, the other fishermen began to notice that Trang had stopped following the rest of the fleet and headed in his own direction every morning. He no longer showed off how much fish he had caught at the end of the day, yet he and his crew seemed to be doing well. Thinking Trang had found a secret fishing spot that he was keeping quiet about, other captains began to follow his boat. There was nothing Trang could do to stop them, and one by one more boats joined in the feast. The area was soon being crisscrossed by more than a dozen sampans, each loading its hull with pottery until it could hold no more. Huge dishes, ornate dragon sculptures, and delicate vases were picked out of the nets and sold straight to the dealers.

After months of intensive trawling, the boats began to catch fewer and fewer pieces. The dealers were paying ever-higher prices, and the

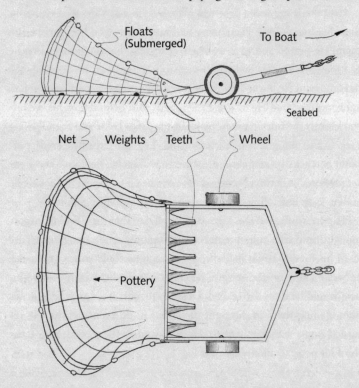

fishermen became increasingly aggressive with each other. The old ca-maraderie of the fishing fleet was forgotten as they scrabbled for the treasure from below. Crews began to devise more effective ways of pris-ing the ceramics from the seabed. Among these was rolling a wheeled metal rake along the bottom, its prongs gouging up the mud and dump-ing it in a net that dragged behind. Most of the pots were smashed by its teeth, but the ones that survived intact made the trips worthwhile.

Nigel Kerr snapped awake as soon as he heard the key slide into the lock of his hotel room. Kerr had left the Special Forces some three years before, but the conditioning was still sharp. He cursed silently as the door pushed open and two Singaporean girls in miniskirts and smudged makeup tottered into the room. Kerr sank back into his bed and groaned. Early-morning light was streaming through a crack in the cur-tains. It was time to get up. "Good morning, Nigel," said the girls, kick-ing off their high heels before collapsing on their beds. Kerr walked over to the bathroom, followed by the dull-eyed gazes of his roommates. Picking up his bag he left the room, saying good night to the girls as they smoked a last cigarette.

This wasn't exactly how he had imagined things turning out when he left the Legion three years earlier—broke and sharing a room with two hookers. Downstairs in the dining room some of the windows were boarded up, the others so dirty that they made the sunlight appear smoky. Kerr walked over to the canteen and poured himself a cup of coffee, then looked around for a chair that wasn't either broken or oc-cupied. Just one hundred meters off Singapore's main street, Orchard Road, the Mitre Hotel had always been a favorite among commercial divers and other offshore oil workers. The family that ran it was in a long-running court battle over ownership, and as a result it had re-mained run-down and cheap for the last few decades, even as neighbor-ing real estate rocketed. A man called "Uncle" ran it, treating divers the way he'd treat his family—rough but fair. A back room was filled with suitcases left by men who'd gone offshore and never come back, while a

mirror behind the bar was covered in oil company stickers. The breakfast room was starting to fill up, and Kerr watched hungrily as eggs on
toast were delivered to tables. Digging into his pocket, he pulled out two
dog-eared notes. His last two Singaporean dollars.

An Irishman from County Down, Kerr had escaped a life of street-
fighting and unemployment by joining the French Foreign Legion. The
Legion is a tough fighting unit, famous for its policy of accepting anyone, outlaws and criminals included, so long as they pass the fierce selection process. New identities are issued to all new recruits, so the
outfit attracts men who have something to hide and little to lose. Selection might be tough, but once you are in things only get tougher. Orders
are given in French, and the beatings meted out by the sergeants mean
that newcomers have to pick up the language fast. But Kerr had thrived,
becoming an unarmed combat instructor and rising to the rank of *Caporal* in the elite parachute regiment, 2-REP.

After five years of operations all around the world, including a long
African war in Chad, he decided to try something else. Looking around
at what was out there, Kerr's eye had settled on the cover of a magazine.
On it was a commercial diver looking every inch the spaceman, beams
of light from a huge helmet cutting through black water. Inside, the article spoke of the adventures to be had on the underwater frontier, a life
of big risks and big money. With his pay slips from the legion, Kerr had
more than enough cash to invest in getting trained. He could have taken
a three-month commercial diving course, but with the funds and ambition to get properly involved, he signed on to a two-and-a-half-year marine engineering degree course in California. As if to distance himself
from his crew-cut past he grew an afro, while working security on night-
club doors to pay the rent.

Marine engineering had undergone a revolution thanks to Cousteau
and Gagnan's invention. The automatic demand valve that made scuba
possible had consequences far beyond recreational diving, for a new era
of commercial diving was born when the oil industry woke up to the op-

portunities that lay underwater. Cousteau had launched himself as an ambassador of the deep, and along with Arthur C. Clarke, he began advocating the seabed as mankind's next colony.

The main problem for human beings underwater is the difference in pressure between the seabed and sea level. Standing at sea level, the air above you—a column that stretches from your shoulders sixty-two vertical miles into space—squashes everything in and around your body with a pressure of one "atmosphere." It takes only thirty feet of water to do the same job, and as divers go deeper, the pressure mounts fast. At thirty feet, two atmospheres' worth of pressure are acting on you (one from the water, one from the air in the atmosphere); at sixty feet there are three, and so on. At high pressure, the gases you breathe behave differently and the body reacts to them in strange ways: Oxygen, the life-sustaining fifth of the air we breathe on the surface, becomes lethal; nitrogen, the harmless majority at sea level, turns narcotic. Adjusting the proportions of the mix can help, as can replacing the gases causing trouble, such as nitrogen, with those that don't, such as helium. The snag is that even the toughest diver eventually gets tired and needs to eat and sleep, and to do so must return to the surface. And that takes time: Whichever gases have been breathed, the body has to depressurize very slowly to stop bubbles from fizzing out of its tissues.

In 1959, as the first satellites were going into orbit around the earth, Captain George F. Bond of the U.S. Navy had an idea. Unwittingly, he was about to start a race for the deep that would parallel the struggle to reach the stars. It was the challenge of the sea, a grittier, industrial version of its stellar counterpart, that would go on to attract men like Nigel Kerr. When Bond met Cousteau in New York he explained his idea. The tissues of the human body absorb more gas at pressure than they do on the surface, but they don't absorb an infinite amount. After a while the tissues become saturated with gas; they can't hold any more. After this point, staying longer underwater does not mean having to do more decompression. So, suggested Captain Bond, what was needed was an underwater house. Divers could live there, and only when their work was done would they have to decompress—a process that might take several days, depending on the depth at which they'd been working, but that was still preferable to doing so repeatedly. Saturation diving was born, and with it a new breed of commercial diver prepared to live at massive pressures for weeks on end without relief.

In 1962—nine years before Salyut 1, the earliest space station, was built—the first underwater base was established. Cousteau, Bond, and an engineer named Edwin Link designed an undersea habitat sixty feet underwater. This first experimental station was occupied for only fourteen hours, but was the start of an underwater race that paralleled the one for outer space. Cousteau launched the Conshelf program, while Bond started Sealab. Cousteau's Conshelf 2, built in 1963 in the Red Sea, represented the first attempt to start a human colony on the seabed. Researchers lived for a month in Starfish house, the main complex at thirty-six feet deep, while a parrot named Claude acted as a canary—he'd be the first to die if the air went bad. The following year, divers in Bond's Sealab 1 off Bermuda undertook the first open-water saturation dive. It was a revelation: Without the need to decompress after every outing, researchers could spend as long underwater as they needed to. The possibilities for industry became clear in 1965 when oil executives watched televised feeds from Cousteau's Conshelf 3 habitat 328 feet down off the coast of France. Though the French petrochemical industry was part-financing the Conshelf program with an eye to exploiting the seabed, they stared unbelieving as divers attached a four-hundred-pound repair assembly to a wellhead more quickly than land crews could, helped by the relative ease of moving heavy equipment underwater. Cousteau subsequently dropped the Conshelf concept, frightened by the industry's enthusiasm and by the thought of rampant exploitation of the oceans, and turned to marine conservation instead.

The Sealab projects continued. Sealab 3 got under way in 1969, taking fifty divers down to 620 feet off San Clemente, California, while others made excursions to one thousand feet. However, when a helium feed sprang a leak, one of the divers sent to repair it asphyxiated from carbon dioxide and died. The U.S. Navy pulled the plug on Sealab, fearing for the safety of the other divers. It was the end of the deep-water habitat programs.

Fortunately for Kerr and the other residents of the Mitre Hotel, the oil industry wasn't so easily put off. With large proportions of the world's oil and gas reserves known to be underwater, they poured money into making saturation technology practical in offshore industrial applications. They soon realized that habitats did not need to be underwater, and that they would be much more manageable (and cheaper) if they weren't. Divers could be brought to the surface in pressurized containers—diving bells—and unloaded through air locks into chambers on the deck of surface vessels. That way they could remain at seabed pressure for weeks at a time while still returning to the relative safety of the deck.

Commercial saturation technology began in the Gulf of Mexico as the United States started to tap the oil and gas reserves there. It quickly spread to the North Sea, where both the equipment and the nerves of its operators were stretched to their limits. Universally acknowledged to have the toughest offshore underwater conditions, the bone-chilling waters of this region are often churned into raging seas. In those early days there was no training; a diver was expected to learn on the job. There was scant regard for safety, and gruesome accidents were common. Dive-support vessels would lose their position while men were down, dragging the dangling divers through a forest of pylons beneath a rig. Gas welders, thermic lances, high-pressure water-jet cleaners, and hydraulic grinders became lethal in the murky water, as did the heavy loads that swung unseen from cranes on the surface.

But there was much at stake, and the oil companies were willing to pay a lot of money to men ready to take the risk. It was common for a North Sea diver to earn $3,000 a week in the 1970s. An average house in Britain cost less than $10,000. The job attracted men (and a few women) of a certain frontier mentality. After many gruesome deaths (forty-five between 1971 and 1979 in the North Sea alone) the regulations were finally tightened up, and oil companies were forced to comply with stringent health and safety regulations.

Though the challenges and rewards of a career beneath the North Sea were what had beckoned Kerr, at the end of his marine engineering degree, he realized he'd made a small miscalculation. He'd done his diver training in the United States, but the U.S. qualification didn't work in the North Sea, where UK and Norwegian companies used British safety standards. What's more, with no green card he couldn't work in the U.S. oil fields either. Kerr took the one option left open to him and bought a one-way ticket on Aeroflot to Singapore, the gateway to the oil fields of the South China Sea. Along with the Nigerian offshore industry, that of the South China Sea had a reputation for offering opportunities to inexperienced young divers. Comparatively free of regulations and with a pool of recruits eager to get involved, the dive companies could impose harsh work conditions. No one minded; everyone just wanted to get in on the saturation scene.

Looking around the dingy dining room in the Mitre Hotel, it was easy for Kerr to spot the other divers. Those looking for work were up early, ready for another day knocking on doors, while the ones who had just come off jobs were hung over or still asleep after a hard night's partying. They were high-rollers for a while when they came back in; they could afford anything. One story tells of two divers who had just come out of saturation in the North Sea and stopped in London on their way back home. Wandering the streets one morning looking for a drink, they blundered into a wine auction happening at Christie's. They started bidding on a pair of bottles. Unconcerned as the price kept rising through the thousands, they continued to bid until all the other parties dropped out. The two roughnecks walked up to the counter to collect their wine, paid in cash, then asked for a bottle opener and some glasses. Shocked, the attendant suggested to the men that the wine was an investment, not a beverage, and besides, Christie's wasn't licensed to serve drinks. The divers shrugged. "No worries," one said, and smashed the cork off the bottle before starting to swig from its broken neck.

Such extravagance was a long way from Kerr's situation. He faced another day of knocking on dive-company doors. Many ex-forces men had problems adjusting to life on the outside and found themselves lost without the iron discipline they'd become used to, but not Kerr. He knew he'd always be okay. Perhaps it was because the army had never "broken" him like it had so many others. There hadn't been the need. He'd been able to take the orders and carry them out easily, and he'd come out smiling. Others had written books about how hellish their experiences in the Legion had been, but for Kerr it had been fun. His only regret was that he hadn't seen as much combat as he'd have liked. He drained his coffee and was heading out when the hotel's receptionist called him over to the phone. It was a salvager named Dorian Ball. He'd heard about Kerr through the grapevine: The Irishman's pavement-bashing had finally paid off.

CHAPTER 4

A Trial Run

DORIAN BALL HAD BEEN a diver on Hatcher's salvage of the *Gelder-malsen*. He had been so inspired by the money that the auction had raised that he started searching for a treasure wreck of his own. After some time he came upon a ship called the *Diana*, which had been trading on an English East India Company license between China and India in the early nineteenth century. She had carried cotton and opium to the Chinese, and returned with silk and porcelain to sell in India. According to the archives, the *Diana* disappeared while passing through the Straits of Malacca one night in 1817. When Ball rediscovered the ship he began negotiating with the Malaysian government, and eventually won rights to salvage her remains.

AS DORIAN BALL and his team—including Nigel Kerr—began to recover the cargo of the *Diana*, the Malaysian businessman Ong Soo Hin was considering the prospect of a different wreck in the Malaccan Straits. In the post-*Geldermalsen* fallout, Ong had seen a chance to make money out of shipwrecks legally. The far stricter conditions many governments were now imposing required that any recovery from an old vessel be carried out archeologically, not simply as salvage. Ong had, with typical thoroughness, looked closely at what was involved in archeology and decided that it was, in essence, a very controlled recovery. He had read in several places that the object of underwater fieldwork was to bring up the remains of the ship in such a way that it could be reconstructed accurately when back on the surface—in a virtual sense if not in physical reality. All that was required, Ong deduced, were carefully thought-out procedures. So long as everything was recorded accurately as artifacts were lifted off the site, the scientists would be happy and his salvage would be not only legitimate but laudable. For all the merits of his idea, Ong's rough-and-ready partner Hatcher was unwilling to consider it. The two men worked together on only one more salvage, that of a World War II merchant ship with a cargo of tin that had sunk off Vietnam. Ong negotiated with the Vietnamese Salvage Agency and its dapper director, Hoang Van Loc, while Hatcher ran the operation. Though everything on this project was done with the correct permissions, the difference between Ong and Hatcher's views over legality appeared to ferment.

Ong's partnership with Hatcher, which had lasted for nearly ten years, eventually ended in acrimony. Whether Ong's new ideas were to blame or whether things had started to go wrong elsewhere is hard to ascertain. Versions of the story involve a contract broken and an investment gone wrong. Most alarming of all, soon after he and Hatcher parted company, Ong was attacked outside a restaurant in Singapore by a hired thug wielding an ax handle. The assault left his jaw badly broken and it had to be wired in place for a year. Ong took his bitterness to the gym, starting an obsession with fitness and muscle tone. As with his yacht, his Austin-Healey, and his tennis, Ong strove for mastery, in this case of his own physiology.

While his bones set and his strength grew, so did his determination. Though he had rarely been on board one of his ships, he had watched how Hatcher operated. Ong now not only knew the business side of salvage and had access to its finance, he understood the techniques involved. Why should he share the profits of his next project? The careful recording he was proposing would take far longer than Hatcher's smash-and-grab approach and so be more expensive, Ong realized. But when compared with the costs of confiscated cargoes, impounded vessels, and court cases, the legal route promised a better yield. Moreover, the relationship between archeology and salvage could be mutually beneficial rather than a detrimental compromise. During the *Geldermalsen* sale the story of the wreck had been as important in the success of the auction as the quality of the artifacts, and marine archeologists were in the business of uncovering the stories that were hidden in a shipwreck. By doing their work, they would help Ong establish valuable provenance. In return, he could help them reach deeper wrecks with technology otherwise beyond their limited funds.

First, however, Ong needed to establish his credentials as an ethical salvager and get out from under the shadow of the *Geldermalsen*. Before his falling out with Ong, Hatcher had come across the remains of the Dutch East Indiaman *Nassau* in the Straits of Malacca. He had dug around and recovered an astrolabe, a valuable navigational instrument, and some breech-loading bronze cannon, but, finding no rich cargo, he had moved on. Ong now realized that what the *Nassau* lacked in treasure it made up for in archeological interest. Though the valuables were gone, the ship's story remained. The *Nassau* had caught fire and sunk during the Battle of Cape Rachado in 1606. The naval confrontation had been a pivotal point in Malaysia's history, marking the start of the Dutch supremacy over the Portuguese. Ong convinced the National Museum of Malaysia that the ship's remains would make a popular exhibition, and the government agreed to pay for the work once it had been done—all Ong needed to do was organize the initial investment. Given that the *Nassau* had no treasure on board, no one could accuse him of seeking it, and by pulling off a bona fide excavation of the ship he would be able to convince the governments of Southeast Asia of his conversion from treasure hunting to doing philanthropic work. The one thing Ong most needed was a prestigious archeological director.

MENSUN BOUND, editor of the *International Maritime Archeology Series,* director of Oxford University's Maritime Archaeological unit, veteran of more than twenty excavations, and presenter of the Discovery TV series *Lost Ships,* saw the offer as a godsend. Not only would a project in Southeast Asia open up a whole new area of the world to him, partnership with Ong would relieve him of all the pressures of fundraising and logistics, permitting him to concentrate on the archeology.

The only sticking point concerned the politics of archeology at the university. Ever since the rebuff Mensun had received from the receptionist on his first day in Oxford he'd felt unwelcome at the Institute of Archaeology. When MARE had been set up, its board had proposed establishing an office at the institute but was turned down. Mensun was not even granted a key to the resource room, a courtesy usually extended to every graduate student. MARE operations had to be sanctioned by MARE's board, on which important members of the Institute of Archaeology sat. When they heard of the *Nassau* project and learned of Ong's involvement with the *Geldermalsen,* they began to kick up a stink. John Boardman—now Sir John Boardman—put up a staunch defense, however, and was assisted by the chairman of MARE's board, Lord Bullock. A modern historian, Bullock had extensive experience both working in a fledgling academic field (modern history was not considered a worthy subject at the university until he made it one) and raising funds (he had founded his own Oxford college, St. Catherine's—a remarkable financial feat). The chairman appreciated the importance of good financing, and since there was no commercial involvement in the proposed excavation of the *Nassau* and therefore no ethical problem, he convinced the rest of the board to accept Ong's proposal. Mensun was thrilled. Not only would the *Nassau* be the oldest Dutch East Indiaman to be examined by archeologists, she also had been involved in events that shaped the course of Southeast Asia and influenced the balance of power within Europe.

THE *NASSAU* was the first salvage project that Ong put together on his own, without Hatcher to run the operation at sea. Rather than relinquish control by bringing in another partner with salvage experience, Ong decided to hire his own experts. While he was still in partnership with Hatcher, a self-assured young Irishman with a degree in marine engineering had turned up at the dockyard one day looking for work. They hadn't been able to take him on immediately, but when hiring his team for the *Nassau,* Ong had dug out the man's contact details. When Nigel Kerr came in for an interview and Ong heard more about his background as an unarmed combat instructor in the Legion, he was immediately hired. Though his official task was to help put together a purpose-designed excavation barge for the *Nassau* project, Kerr soon found himself acting as Ong's right-hand man, working by his side for most of the day and eating with him in the evening.

Used to the tight budgets of university projects, working with Ong on the *Nassau* was a revelation for Mensun. The *Abex TS,* the vessel that Kerr and others were putting together, allowed his laboratories to be housed directly above the site and his divers to work with commercial underwater apparatus. Though the wreck was only one hundred feet down, the Straits of Malacca were not an easy place to work. The waters were murky and the currents strong, conditions that could easily lead to a scuba diver being swept away. Having his divers connected to the surface via their umbilical air hose (as the commercial diving equipment required) reduced the danger somewhat, though not entirely. At the start of a sortie each diver would secure their umbilical to the dive stage so that if they were carried away by the current they could haul themselves back into position on the seabed. It wasn't a standard procedure, for if there was a problem the deck crew would not be able to recover the diver. But it was preferable to the alternative. One day a diver forgot to tie himself to the stage and was picked up by the current. As he flew through the black water he began to rise, swinging upward on his umbilical as it pivoted from the front of the barge. Only his screaming prevented his lungs from bursting during the rapid decompression, and seconds later he appeared on the surface behind the barge. He was eventually hauled onto deck and rushed into a pressure chamber that was kept on board for such emergencies, and survived any further injury.

The MARE divers who had worked on the wreck returned to the UK energized and enthusiastic. I'd been on a different MARE expedition— to Uruguay, on the other side of the world—and was envious when I heard all that I'd missed. Mensun felt he had graduated to a new league of operation, and when he published his excavation report he concluded that "the excavation of the *Nassau* was the largest and most expensive maritime archaeological project that the world has seen in recent years." The story the team uncovered was dramatic, for the evidence showed that the *Nassau*'s demise had been violent. She had taken a beating from the Portuguese: Impacted cannon and musket balls riddled her timbers. It seemed that by the end, the *Nassau* had been defenseless—only thirteen cannonballs remained in her stores. Charred deck beams, cracked iron cannon, and melted bronze guns all attested to the fire that had finally sunk her. A single eyewitness report of an explosion in the powder magazine was also borne out: Her stern had been ripped apart, scattering artifacts far across the seabed.

Ong was also pleased with the results of the project. He hadn't spent much time on board the excavation barge himself; he found it difficult to work without proper communications, and the motion in bad weather made him uneasy. Nevertheless, it was his first solo salvage project, and through his meticulous planning everything had gone smoothly. The sea might be unpredictable, but it could still be bent to a budget. His trial run had worked; his reputation as an archeological facilitator was ensured. Now Ong needed to find a wreck that was inaccessible to salvagers without archeological credentials, a wreck with a cargo that would make him rich.

CHAPTER 5

The Grinning Mandarin

GLIDING OUT OF the Asian history section of the National Museum of Malaysia, hands folded across his belly, Ivo Vasiljev wore a hint of a smile on his face. Like a Buddha, the sixty-year-old Czech linguist remained serene despite his situation. Though blessed with an academic brain that had absorbed twelve different languages to a level unattained by most native speakers, Ivo had been forced to take work for a big, faceless management consultancy that had posted him in Bangkok, far from his wife and children. A professor's salary in Eastern Europe could not support a family, and other jobs in academia were hard to come by. Ivo was making lots of trips—this one to Kuala Lumpur to advise a big American company—but after a varied life of intellectual challenges, his soul was withering in the world of business. Sundays were the only days that he could air his mind, and this time Ivo had decided to indulge his passion for the past.

Settling down in the museum café for a cup of coffee, Ivo opened up a copy of the *New Straits Times,* Malaysia's biggest-selling newspaper. Inside, he started reading about a recent shipwreck excavation in the Malacca

Straits: that of the *Nassau*. Aside from the story of European ships bat-
tling it out for control of the East, it was the idea of collaboration between the government, private sponsors, and the museum that caught
Ivo's attention. The project was a novel way of financing an academic
endeavor. He decided to write his own version of the story for the small
Czech oriental studies magazine *Novy Orient,* flagging the excavation as a
possible model for funding future scholarly work. It was a minor decision to write the article, taken on a whim. Ivo liked to contribute to the
magazine occasionally; it had been running since 1945, and was considered something of a national treasure. He couldn't have foreseen that
penning the piece would completely change his life.

NOT ALL COLLABORATIONS between the National Museum of Malaysia
and salvage companies went as smoothly as the *Nassau* excavation that
Ivo so admired. When the Malaysian government awarded Dorian Ball
the contract to excavate the remains of the *Diana,* they laid down strict
terms about how the excavation should be controlled and the manner in
which an archeologist would supervise. Unfortunately, when the job was
finished the government decided that those terms had not been followed, and so paid him less than the amount that their contract had
specified. Outraged, Ball had taken the government to court.

In order to persuade the court that Ball's conduct on the wreck had
been less than archeologically sound, the government needed to bring
in an expert witness. Already well known to the National Museum of
Malaysia after the *Nassau* excavation, and experienced at working in the
same waters in which the *Diana* lay, Oxford's Mensun Bound was the obvious choice to present the government's perspective. He later recounted
to me how the trial had unfolded.

One of the pillars of the government's case was the young archeologist who had been assigned by the museum to supervise the excavation.
Indeed, it was likely his reports that alerted them to Ball's conduct in the
first place. The waters of the Malacca Straits can be pitch black and the
currents run hard. It is a difficult environment even for an experienced

diver. The supervisor had dived only on eastern Malaysian shores before, in clear waters on coral reefs, and never pretended otherwise. But on the first day Ball had sent him down into the swirling darkness forty meters below on his own. It was, according to the young man, a terrifying ordeal. As a result, the archeologist rarely visited the seabed, and so was unable to check what was happening there. The work became, in Mensun's words, "a smash-and-grab job . . . they just rushed everything up to the surface any way they could," techniques that Ball had presumably learned from Hatcher during the *Geldermalsen* salvage. Mensun also showed the court photographs that had been taken of the seabed, explaining how a site should look and how this one differed. He proved there had been no recording of where the artifacts had come from. He went on to show that the site-plan diagram had been fabricated. Mensun even denounced the book that Ball had produced on the wreck, telling the court that it was not an archeologically credible publication. The court awarded the case to the government. Mensun's reputation as a bastion of archeological standards in Southeast Asia was assured.

BACK IN OXFORD, a few months after returning from the *Nassau* excavation, Mensun received an unexpected invitation from the university's geoscience faculty, asking him to look at something that had recently come in. At first he tried to decline the invitation, but his colleague from the Thermoluminescence Laboratory was insistent. Mensun would be interested, she assured him.

Thermoluminesence testing is a valuable tool for archeologists, capable of dating a ceramic artifact (as long as it is between ten and 230,000 years old) by revealing how long ago it was fired. Clay absorbs radiation, but the heat of a kiln during the manufacture of a pot burns it out, effectively starting a stopwatch as the ceramic begins absorbing afresh. When a sample is reheated to 500 degrees Centigrade in the lab, it glows. Measuring how much it glows gives a precise indication of how much radiation has been absorbed, and therefore how many years have passed since its firing.

The geoscience faculty was housed in a maze of low corridors in which Mensun soon got lost, and he began to wish he had got the better of his curiosity. When at last he found the right room, he was shown a slightly dilapidated-looking statuette. The figure was on one knee, with a vase balanced on the other. It was certainly unusual. The pose was familiar from Chinese porcelains and suggested the figure represented a court official, a mandarin. But this wasn't of Chinese manufacture. The clay was heavy and coarse, more stoneware than porcelain. The base-glaze was stained in places and creamy rather than white. The blue clouds on the figure's robe were lighter than the blue swirls on his hat, as if the pigments had been inconsistent. The work was definitely Southeast Asian, heavily influenced by Chinese styles of the fourteenth or fifteenth century. It wasn't Thai, Khmer, or Burmese; it was far too sophisticated. Mensun's eyes followed the blue lines traced around the statuette's mouth. Toward one corner of the top lip, the painter had departed from the contours of the molding with a subtle twist. That wouldn't have passed quality control on the Chinese production lines. Then he noticed that the right eye was narrowed, too. Neither irregularity appeared to be a slip. Together, the eye and the mouth gave the mandarin an amused grin that crept up the right side of his face. He might be on his knees, but he wasn't cowed. Mensun couldn't help grinning back.

As was often the case with antiquities that the lab had been asked to date, the technicians had signed a confidentiality agreement and couldn't tell Mensun who had sent it for analysis or why. Thermoluminescence testing was expensive, and usually requested by a wealthy collector or an auction house wanting to test the authenticity of a valuable piece. Was someone worried the mandarin was a fake? Mensun doubted that it was. A forger would not have chosen such an unusual style. It was possible the owner was simply trying to ascertain in what period it had been made, but sending it to the lab would be a last resort; thermoluminesence had a margin of error of 15 percent, or about seventy-five years, in this context. For most styles of pottery, such as Chinese, the style of painting provided a more accurate dating method.

Chinese porcelain had been studied and admired for centuries. Sophisticated refining techniques, removing iron from the water, combined with elements in the Chinese soil—kaolin and feldspar—made the clay fuse seamlessly with its glaze. This vitrification produced the delicate white porcelain that was treasured all over the world since well before A.D. 1000. No one else could figure out how it was made, and the Chinese weren't going to reveal their secret. When the first examples started reaching Europe, they were thought to have been made from conch and eggshells that had been ground up and mixed into a paste with egg white then buried in soil for eighty to one hundred years and only then judged to be seasoned and ready for shaping, decoration, and final glazing. This myth persisted in Europe for at least fifty years, until the mid-sixteenth century. The actual production process was tightly controlled: Paint pigments were manufactured with the best-quality cobalts, and the artists were so rigidly schooled in the traditions of their craft that motifs and designs varied little over decades, sometimes even over centuries. Successive Chinese dynasties kept beautiful collections of the highest-quality imperial wares together, and a vast amount had been produced. In the sage words of Tomé Pires, a pharmacist based at Malacca writing in 1515, Chinese porcelains were "truly infinite, and therefore too tedious to detail." Nearly five hundred years later, so much had been written about Chinese porcelain and in such detail that as an academic field it became saturated, with every shift in style and influence analyzed many times over.

The only ceramics that Mensun knew of that came close to fitting the style of the statuette were Vietnamese. While in Malaysia for the *Diana* trial, Mensun had visited the National Museum in Kuala Lumpur, where Ivo had first read about the *Nassau* wreck. Like Ivo before him, Mensun had paused for a long while in front of a display cabinet containing a few pieces of Vietnamese ceramics. They were not true porcelain, for they had been made with the water found in northern Vietnam's Red River delta, water that was high in iron oxide (hence, the color in the river's name). The result was that they had to be made thicker than porcelain to attain the same strength, and so felt closer to stoneware. The pieces were of a higher level of artistic sophistication than the other non-Chinese wares and so stuck in his mind. Then, Mensun had begun to notice examples elsewhere—in antiques shops, particularly—and wondered how he had not paid attention to these pots before.

When he turned the statuette of the mandarin over in his hand, Mensun realized why he had been invited to examine it. There were barnacles on the underside. The piece was from a shipwreck. Perhaps a new site had been discovered, he thought. It would explain the recent trickle of Vietnamese pottery into the antiques dealerships. After leaving the lab, Mensun did some more research and realized that he wasn't alone in having neglected Vietnamese ceramics. Very few people knew anything about them. The few Vietnamese pots in the world's collections were something of a mystery. Occasional pieces showed that Vietnam could produce ceramics more sophisticated than those from anywhere else in Southeast Asia, but most collectors still saw them as poor-quality imitations of Chinese porcelains. Given the examples available for study until now, Mensun would have had to agree, but there was a character to the statuette of the mandarin that intrigued him; it hinted that there was something more to Vietnamese ceramics.

The lack of prior academic knowledge was like a whiff of pheromone to Mensun. Had he no other commitments, he would have reacted immediately. Instead, he forced himself to restrain his instincts. Two MARE projects, a survey of Nelson's favorite ship, HMS *Agamemnon,* off the coast of South America and the excavation of an unidentified Elizabethan warship in the waters of the Channel Islands between England and France, were both still in progress. Besides, knowing the mandarin came from the sea was little help. Even if the statuette and the other pots he had seen were all from a single wreck, Mensun had no way of knowing where in the vast eastern oceans that wreck might lie.

CHAPTER 6

Mensun's Dilemma

ALTHOUGH VARIOUS COLLECTORS, dealers, and academics around the world had noticed an influx of strangely high-quality Vietnamese pottery, no one in a position of authority knew where it was coming from. Then, one morning at Da Nang Airport in central Vietnam, two Japanese art dealers were stopped at customs and each was found to be carrying a briefcase packed not with paperwork or even banknotes, but antique pottery. The pair were immediately arrested for smuggling illicit antiquities. Authorities traced their trail down the coast back to Hoi An, about an hour's drive south. There the investigators were surprised to find that the town's antiques and souvenir shops were full of bowls, jarlets, and dishes of a similar style. Some of the pieces had clearly come from the sea, so the police began to search the surrounding riverside and coastal villages. The fishermen responsible were not hard to identify. Next to the usual piles of ropes, floats, and nets lay weighted metal buckets and long, toothed rakes, and rather than returning with mackerel, sardines, croaker, and squid, their hulls were full of ceramics.

As it turned out, the Ministry of Culture already knew there was a problem somewhere in the area. Dr. Trinh Cao Tuong from the Vietnamese Institute of Archeology had alerted them after he had passed through Hoi An and recognized the ancient ceramics being sold there. They matched "wasters" (pieces that had not survived firing in the kiln) that he had found during excavations at a village in the Red River delta called Chu Dau.

Chu Dau was perhaps the most important archeological discovery in twentieth-century Vietnamese history. Since the 1950s archeologists had been working in Vietnam under the auspices of the Ministry of Culture, Hanoi University, and the Institute of Archeology to start patching together the nation's past from what tatters remained after the war, but Chu Dau was rediscovered by chance. In 1983, a senior Japanese diplomat at the embassy in Hanoi happened to be visiting Turkey and took a trip to the Topkapi Saray in Istanbul, home to the Ottoman sultans from the mid-fifteenth to the mid-nineteenth century. An amateur historian, he was admiring the palace's impressive collection of Chinese blue-and-white wares when he noticed a vase decorated in a very different manner. Though the motifs on it were Chinese, the piece was painted with a freedom that appealed to his Japanese taste for improvised ceramic style. There was an inscription in Chinese characters around its shoulder reading, "In the 8th year of Thai Hoa (1450), in the Nam Sach chau, a crafts(wo)man of the Bui family painted for his/her pleasure." The Japanese diplomat immediately sent a letter to the Communist Party secretary of Hai Duong province in North Vietnam suggesting that they should look for remains of ancient potteries in their Nam Sach district.

When archeologists arrived in the area that same year, they found that in some villages the dirt yards behind the houses were studded with the remains of pots that had collapsed, exploded, or been otherwise damaged while being fired five hundred years earlier. The wasters led to the identification of fourteen ceramic centers, including Chu Dau, My Xa, and Ngoi, all located on the plains of the Kinh Thay and Thai Binh rivers, where both kaolin and fine clay could be easily found. Evidently the Nam Sach district, located at the heart of the Red River delta, had been the center of the country's ceramics industry. Here were the villages that had created the finest artworks in Vietnam's history. The excavation of Chu Dau, the most extensive of the sites, still continues today. Its discovery represented a great advance in Vietnam's understanding of its history, but all that remained of the masterpieces Chu Dau had once produced were the deformed rejects, the magnificent Topkapi Saray vase, and a few scattered examples recovered from Indonesia, Malaysia, and the Philippines. The fruits of the mid-fifteenth century, when Vietnam was at the height of its golden age and the artisans of Chu Dau were at their most accomplished, seemed to have been lost forever.

Dr. Trinh Cao Tuong had recognized immediately that the ceramics in the market stalls of Hoi An were the same as those from Chu Dau, and he warned the Ministry of Culture that an important part of the country's heritage was being lost. However, officials at the ministry were unsure what to do. Their few inspection divers were trained only in shallow inshore and harbor work; this site was in open water. In the end, the small museum in Hoi An was the only institution that took positive action. Its curator reasoned that there was no other way the pottery could be raised than by the fishermen, so he began buying up as much as he could afford. The fishermen knew they could get better prices elsewhere, often selling to dealers in Saigon willing to pay up to ten times the amount they would get in Hoi An. Soon the museum's storerooms were filling up with only the damaged or low-quality pieces that the dealers didn't want.

But with the arrest of the Japanese dealers there was evidence of an international trade, and the prime minister of Vietnam himself took notice, spurring the government into action. Boats were banned from the area, but the authorities knew that they would not be able to guard it forever and that fishermen would eventually begin creeping back to the site. Somehow the shipwreck, or at least its cargo, would have to be recovered.

The Vietnamese Salvage Agency (VISAL) was one of the first groups contacted. Their international operations director, Hoang Van Loc, had experience with historic shipwreck cargoes, for in 1992 the agency had collaborated in salvaging the "Vung Tau" wreck, named after the southern coastal city near which the ship had sunk in the late seventeenth century. Together with a foreign partner, they had managed to raise $7.2 million at auction for the Chinese porcelain that had been on board. More recently, Loc had collaborated with Ong Soo Hin and Michael Hatcher on the recovery of a cargo of tin from a World War II wreck.

This new find was a different proposition, however. The artifacts had not been made in China but in Vietnam and the government was attentive to the fact that the ceramics were a vital part of the country's heritage. The Ministry of Culture decreed that any recovery operation had to be an archeological excavation, not a salvage job. It was a difficult position for Loc, for the area in which the fishermen had been dragging their nets was located in deep water. He knew that operating there would be beyond his agency's budget, and beyond the Institute of Archeology's expertise.

Ong was perfectly prepared for this opportunity when Loc called him for advice. He reminded Loc that he had experience with both the commercial side of historic salvage through the *Geldermalsen* and the academic side, working with archeologists on the *Nassau*. Knowing that every foreign venture in Vietnam required taking on a local partner, he suggested a collaboration between his salvage company (Saga Horizon) and VISAL. Together they could resolve the Vietnamese government's dilemma by carrying out an archeological excavation of the site in return for the right to sell a proportion of the pottery recovered. The government would receive a full archeological report, recover most of their sunken heritage, and also have enough artifacts to fill all their museums. What's more, they might even make some money. All they needed to convince the government to grant them the license to work was an archeologist who would break rank with the academic establishment and participate in a commercial venture. It would not be easy—only an archeologist in a prestigious position would do, yet those who reached such positions were usually anxious to hold on to them.

IVO VASILJEV'S ARTICLE on the *Nassau* excavation was published in the Czech magazine *Novy Orient* in late spring of 1996. Out of courtesy to the Malaysian National History Museum, he arranged to send them a copy of the piece via a friend who was going to Malaysia to visit his family. As an afterthought, Ivo gave his friend, Long Chai Meng, a second copy to send to Mr. Ong Soo Hin, should he be able to find the man's address. In the process of tracking down Ong's whereabouts, Long ended up speaking to him on the phone. While explaining about the article that he'd been asked to send, Long happened to mention that its author was a scholar of Vietnamese language and history. When Long returned to the Czech republic he described to Ivo how pleasant Mr. Ong had been on the phone, and how delighted he was to have been contacted. He had even expressed interest in meeting Ivo the next time he was in Kuala Lumpur. Ivo, unaware that Long had mentioned his Vietnamese affiliations, thought that Ong was simply pleased to have received a mention in the Czech press.

About a month later, while in Kuala Lumpur for a regional training weekend, Ivo took Ong up on his offer and arranged to meet in Ong's office in Petaling Jaya, on the outskirts of the capital. When Ong invited him inside, Ivo was surprised at how small the premises were. Given the press, he'd imagined something much bigger. However, when the two men began talking about the *Nassau* and the mix of business and scholarship, Ivo found himself impressed by Ong's clear thinking. Then Ong asked if it was true that Ivo spoke fluent Vietnamese. When Ivo confirmed that he did, Ong came clean. He was about to start a project with an ancient ship in Vietnam, he explained. Nothing was confirmed yet, but if the excavation went ahead, would Ivo be interested in joining the team?

Ivo's heart leaped, then fell just as fast. Nothing would give him more pleasure than to be involved in a project such as the one about which he had written his article, especially if there was a link to Vietnam. But, he explained to Ong, with his financial situation there was no way that he could leave his job for just a few weeks. He could only consider a long-term position. Ong nodded, and said that there was a good possibility. He would get in touch if the project went ahead.

At the beginning of 1997, Ivo finally received a fax. After months of visits, negotiations, and endless committee meetings, the Vietnamese government had granted Ong a license to search for the wreck.

NIGEL KERR had been chief diver on the *Nassau* excavation, and so Ong asked him to go to Vietnam in April 1997 to locate the wreck off Hoi An. After some encouragement from the authorities, the fishermen agreed to lead Kerr's search team to the area where they trawled for pottery. However, each day a light mist hung over the water, obscuring the distant mountaintops and island hills the fishermen used as navigation guides. For the whole first week, Kerr was thwarted.

When the mist finally dissipated, Kerr plotted an initial search zone of five square miles. Once this was done, he began to do some trawling of his own, using a sonar "fish" in place of the fishermen's rakes. For over a week he squinted intently at the monitor, shielding it from sunlight with his hands as he sat propped up against the sides of the boat's cabin. All day every day he watched the twin bands of gray shading track slowly down the screen, moving only when the boat turned at the end of each search line, at which point he would switch his position to the other side of the cabin. Kerr was accustomed to a little more action than merely moving out of the way of the sun but forced back his impatience, and tried to ignore how slowly the serrated contour of the mountains on the distant mainland crawled past.

Action was what the Foreign Legionnaires lived for, and action was what they'd got in that same Vietnamese mountain range forty-three years earlier, at the battle of Dien Bien Phu. One of the most disastrous defeats in French military history, the fight was also a characteristic display of the Legion's tenacious attitude. The French had occupied Indochina (as it was then known) since 1891, but the Vietnamese resistance hadn't taken long to boil up. In 1930 a young scholar called Ho Chi Minh founded a resistance movement that seized back power in the aftermath of World War II and declared Vietnam an independent nation once more.

France had different ideas. Vietnam had been profitable to them, and just six months later they began trying to reassert their rule. Badly weakened by the World War, they soon needed help in the fight, and found it first from Britain and later from the United States. But the Vietnamese would not give up easily and the war dragged on. By 1954 the Americans were paying 80 percent of its costs. Then came Dien Bien Phu.

In the first of a series of blunders, the French general Henri Navarre had decided to build a base in a valley deep in the mountains of Vietnam's northwestern corner. Though it was at the extreme limit for the air support on which it depended, the location was a strategic crossroads between Vietnam, Laos, and China. General Navarre judged the forested hillsides impassable to heavy armor and assumed few Vietnamese were hidden among them. From the hills the Vietminh, under General Vo Nguyen Giap, watched the French base being built and knew exactly where each defensive position was located. Giap compared the location to a rice bowl and began hauling heavy artillery through the jungle to its rim, where it overlooked the French in the valley below. The Vietminh spent months camouflaging gun emplacements, stockpiling ammunition, establishing anti-aircraft posts, and sending spies into the French camp to discover the positions, calibers, and numbers of all their enemy's guns.

When the Vietnamese artillery opened fire for the first time, on March 13, 1954, the French were taken totally by surprise. Giap had ordered decoy gun emplacements to be built complete with wooden guns that would draw the French fire. By firing only a few salvos from each real gun before the next emplacement began the deception was complete. Whereas Giap knew exactly how many guns the French had in each location, the French had no idea where to start their retaliation. The French artillery commander, panic-stricken at his situation, com-

mitted suicide with a hand grenade in his bunker. When the first
French reinforcements appeared in the sky and the Vietminh anti-
aircraft guns opened up, the French realized they were in trouble.
Though they had more than ten thousand troops in the base and would
land over sixteen thousand more, the Vietminh had more than
100,000 men in the hills surrounding them. When General Giap realized that the French were weakening, the battle became a siege. He
stopped sending his men in massed charges and started a steady attri-
tion that he called "slowly bleeding the dying elephant." With the aircraft runways inoperable, the French could only resupply by parachute,
and half of the drops landed on Vietminh-controlled ground. It was
into this desperate situation that the last reinforcements were sent. Between March 14 and May 6, French Foreign Legionnaires from the
third and fifth battalions parachuted into the besieged strip of jungle
at night, their way lit only by the flashes from the gun emplacements.
The four thousand troops did not last long. The final Vietnamese assault took place on May 7, when the remaining three thousand French
troops still guarding the headquarters were overrun by some 25,000
Vietminh, whereupon they finally surrendered.

After the disaster at Dien Bien Phu, the French were forced to agree
on a compromise: Ho Chi Minh's Communist Party could rule the country north of the seventeenth parallel and the south was left to the French-
sponsored forces, from which the Republic of Vietnam evolved a year
later. The peace was fragile, for the United States was convinced that if
South Vietnam fell to the North, communism would begin to spread
throughout the world, while in North Vietnam and among southern dis-
sidents, resentment against the continued American-backed occupation
grew.

In 1963, the situation on both sides of the seventeenth parallel
started to combust—literally. In South Vietnam, Buddhist monks set
themselves on fire in protest at the occupation, and in 1964 the United
States began bombing strategic targets in the North. The Vietnam War
was brewing, but like the Chinese before them, the Americans learned

that whenever Vietnamese forces were stamped out in one place, more sprang up elsewhere. Operation "Sea Dragon," begun in 1966, was aimed at stemming the flow of seagoing junks full of arms, supplies, and men that were streaming down the coast and supplying resistance forces to the south of the seventeenth parallel. By 1967 the campaign had succeeded, but the North Vietnamese simply switched to sending supplies by the Ho Chi Minh Trail instead.

The United States ultimately sacrificed more than 52,000 lives, pouring seven million tons of explosives and $170 billion into the war, but in the end the world's most powerful nation was worn down by the strength of Vietnamese nationalism and their guerrilla-style fighting. In 1975, the Vietnamese had won back their independence. The end of the war with America did not signal the start of peace for Vietnam, however. Decades of foreign-backed influence had torn the nation in two, and the northerners now sought to rejoin the halves of the nation in a brutal leveling. The skilled and educated southerners had for the most part worked with the French and American-backed authorities, and it was they who had the most to fear from the retributions meted out by the newly formed Socialist Republic of Vietnam. Already more than half of South Vietnam's population of twenty million had been displaced by the fighting, and afterwards the upheaval continued. Land reforms stripped them of their property, "reeducation" in concentration camps tried to purge them of their Western ideals, and grinding poverty set in. Millions fled the country either on boat or on foot, forcing the United Nations to set up refugee camps in neighboring countries.

KERR GLANCED UP from the monitor and around the boat. Out of the twenty or so people on board, the only ones awake apart from himself were the helmsman, his friend Jon Street, and Nguyen Huu Tai. Representing VISAL, Tai was the only person of any use in Kerr's opinion. Tai had worked with the salvage outfit for twenty years, and yet, despite his aptitude and long service, he had never risen to the ranks of higher management. He appeared cursed by his past. Before the Vietnam War, Tai had worked for an American shipping company in Saigon. His family had been airlifted out, but Tai was left behind, and so endured five years of reeducation camps and communist indoctrination. The experience left him edgy, temperamental, and manic at times, but his professionalism and experience at sea were undeniable.

The others snoring on the boat were all good Communist Party members. Each of the groups represented on the excavation committee had sent delegates to the survey: the National History Museum in Hanoi, the Ministry of Defense, the Ministry of the Interior, the Institute of Archeology, the Quang Nam Provincial Authority for Culture and Information, the Quang Nam Museum, the Museum of the Revolution in Ho Chi Minh City, the Center for the Management of Historical Monuments of the Hoi An Municipality, and, of course, the Ministry of Culture. Kerr wasn't surprised they were all sleeping. None of them was doing anything except waiting for him to announce that he had found the wreck. All except the secret policeman, that is, who watched every movement the foreigners made with his one good eye.

Tai wasn't spared the eyeball either. Able to speak fairly good English and get things done, Tai had acted as interpreter when Kerr had first gone out with the fishermen to mark out the search area. It was also Tai who had arranged for this boat, borrowing it from the Hoi An District Sea Swallow Nest Exploitation Unit (which collected the nests formed by the sea bird's spittle to sell in restaurants, where they were served as a delicate soup). The thirty-foot wooden sampan was barely big enough to carry the crew, the aforementioned committee members, the sonar, the depth finder, and the undersea robot they would use to investigate potential sites.

Pulses of sound were pinging out of the sonar "fish" that the boat was towing, hitting the seabed in a band that stretched one hundred feet to either side of their path. The "fish" listened for echoes and plotted the results on the screen. Anything standing proud of the bottom cast a sonar shadow, so long as it wasn't in the blind zone directly beneath the

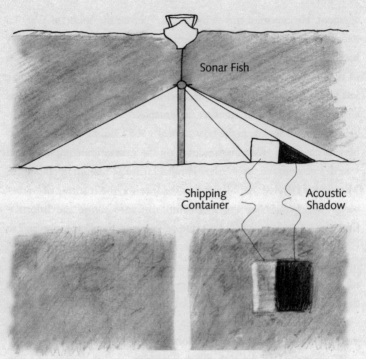

Sonar Fish

Shipping
Container

Acoustic
Shadow

Side-scan Read-out

apparatus itself. What they were looking for would show up merely as a mound, the classic sign of an old shipwreck whose cargo or ballast had slumped as the wooden hull disintegrated. It should be easy to distinguish from the sharp-edged boxes of lost shipping containers that were the only things that had shown up so far.

Close to sunset on the seventh day of scanning, Kerr was beginning to wonder whether he might have to widen the survey zone. Then, just as they were nearing the northern boundary of his search area, the left-hand band on the monitor began to show a gentle rise on the seabed. He didn't say anything at first, letting the skipper keep the boat running in a straight line. Jon Street—another ex-Legionnaire—was bent over a small, hand-held GPS unit, one arm guiding the helmsman's track. The mound kept rising.

"Jon, take a mark on the GPS. Something out to our left there about fifty feet out. Keep the next track nice and tight?"

"Roger," said Street. Kerr had been a *caporal* (a noncommissioned officer) in his Foreign Legion regiment and though they were friends, it was still Kerr who called the shots.

On the return track, the sonar image became clearer. There seemed to be two distinct mounds, each rising six or seven feet above the seabed, about ninety feet apart. Marker buoys were thrown overboard and the coordinates from the GPS noted. It was too late in the day to prepare the three-foot-long Remotely Operated Vehicle (ROV) for a dive, so they made one more pass with the sonar and left.

Ong joined them on board the next morning. As they motored out to the site, Kerr prepared the yellow torpedo-shaped ROV, checking the camera, lights, and thruster motors before Street lowered it over the side by its umbilical cable. Before the robot had even reached the seabed, the screen broke into static. They assumed water had forced its way inside and hauled the robot back on deck immediately, but Kerr could see neither leak nor moisture. A leak would have been easy to repair. Instead, it was likely that the problem lay somewhere along the three hundred feet of cable. There was no way to fix that in a hurry. They had been sitting too long anyway, thought Kerr. Ong had mentioned that physical evidence from the site would help him convince his investors that their funds were being well spent. It was time to get wet.

Two sets of scuba equipment were on board. The depth-sounder indicated the bottom was about 230 feet down. Reading Kerr's thoughts, Tai reminded him that the scuba tanks were filled only with air, not with a deep-diving gas mix. At 215 feet, there is an invisible physiological barrier beyond which the oxygen in normal air becomes toxic to the human body. Normally oxygen makes up about 21 percent of the air we breathe, but as depth and pressure increase, more oxygen molecules are pressed into that 21 percent. At about 165 feet, the 21 percent oxygen in normal air affects the body as though it were 100 percent pure. At 230 feet, the seven atmospheres of pressure are pushing oxygen into the blood at 147 percent, at which level the deleterious effects are felt much more rapidly. Muscles start to twitch, vision becomes distorted, limbs become uncoordinated, and finally convulsions set in.

Nitrogen doesn't change so dramatically from Jekyll to Hyde when put under pressure, but it does become steadily more narcotic. At one hundred feet it produces a pleasurable feeling; you feel as though you've had a couple of drinks. Everything seems that little bit rosier and less serious. The "raptures of the deep," as Cousteau called the condition now known as nitrogen narcosis, accounts for some of the addictive nature of diving. The effects disappear as you ascend, and are easily handled—so long as you're only at one hundred feet. However, every fifty feet beyond that depth feels like drinking a martini on an empty stomach. Rapidly you start to lose balance and sense of direction; the mind becomes stubborn and can fixate on a certain idea, unable to see alternatives. None of these symptoms would be particularly serious on the surface, but at a depth of 230 feet of water with a limited supply of air and the unknown threats of an alien environment, the raptures can rapidly become a siren's deadly song.

We need the oxygen in our air to survive, but we don't need the nitrogen. Realizing this, divers started substituting it with other gases that caused fewer problems. Lighter gases were chosen because they diffuse more readily, and so leave the body's tissues more easily upon decompression. Hydrogen, being the lightest gas known, was an obvious candidate. The French diving company COMEX estimates that dives of over thirty-two hundred feet would be possible using a mix of hydrogen and oxygen, but there's a drawback. At sea level we need to breathe in at least 16 percent oxygen to survive, but if hydrogen is mixed with any more than 4 percent oxygen, the cocktail explodes. As early as 1945, the Swedish engineer Arne Zetterström came up with a solution. Beyond one hundred feet the body can survive on less than the crucial 4 percent oxygen because of the increased partial pressure. Zetterström had enough confidence in his calculations to dive to 525 feet breathing a hydrogen-oxygen mix. Unfortunately, his surface support team made a fatal mistake on his return journey: They neglected to give him any decompression stops as they brought him back up through the last hundred feet. This is the critical zone for decompression, for the difference between one and two atmospheres of pressure is 50 percent more than the difference between two and three, and so on. For that reason, divers ascend slower and slower as they near the surface. Zetterström was brought straight back up and died. Seventeen years later, a dashing young Swiss mathematician named Hannes Keller came up with a new formula for a diving gas that combined oxygen, helium, and nitrogen. After extensive trials, in 1962 Keller made a dive to one thousand feet off the Californian coast. *National Geographic* journalist Peter Small descended with him to record the historic event. Fitting the spirit of the space age, when they reached the seabed Keller left the bell to plant two flags, one for the United States and one for Switzerland. Unfortunately, Keller became tangled in the billowing folds of the Swiss flag. The tank of compressed gas mixture on his back was rapidly exhausted as he fought to extricate himself and he made it back to the bell only by holding his breath. He was so relieved to get inside and shut the hatch that he didn't realize that one of his fins had caught in the seal. When the bell was winched back upward, no one realized that the hatch was leaking. Keller and Small were coming up from their pioneering dive without being properly decompressed. When the men eventually stopped answering the bell telephone, two divers were sent down to investigate. Both men were found unconscious, with the bell still at two hundred feet. The bell was rushed to the surface to get the men medical attention but, though Keller eventually recovered, Small died of explosive decompression. To compound the tragedy, in the panic to retrieve the bell one of the two support divers had been swept away and was later found drowned.

PREPARING FOR HIS DIVE to the seabed off the Vietnamese coast, Kerr dropped a substantial anchor connected to a float on the spot that he'd marked with the buoy. The line would stop him and Street from getting carried away by the currents while also acting as a guide to the mound that he'd seen on the sonar. Losing the line would be fatal, since dangling on it were spare air tanks that the two of them would need to switch to during the long decompression. Kerr and Street looked at each other for a moment. They were in the open ocean, diving to a depth where their breathing gas would be toxic; they had no backup and no chance of getting to a decompression chamber if anything went wrong. They started suiting up, with Tai cackling with laughter at their decision.

Kerr's initial training as a diver had been in the elite *Commando de Renseignement et d'Action Profondeur,* the French Foreign Legion's parachute regiment (since renamed *Groupement de Commandos Parachutistes,* presumably due to the previous title's unfortunate acronym, CRAP). On high-altitude, low-opening (HALO) jumps, they would skydive carrying full kit, often at night, breathing oxygen because of the altitude. Opening the canopy at the last minute they'd hit the water, cut away the parachute, and go straight down, stopping only when they were fifty feet below the surface. They would then begin to swim underwater toward their target.

Though Street wasn't a CRAP veteran, he had done time in the British army's Special Boat Service, the marine version of the Special Air Service (SAS), and had remained extremely fit. Still, it had been a while since he had dived deep on air. As they neared the seabed, Street became disoriented. Kerr was carrying a floodlight that illuminated what appeared to be a field of mushrooms below. The oxygen in the air was becoming toxic with the depth and Street was hallucinating from the nitrogen. Being physically fit only made it worse. He tapped Kerr and indicated that he was going to go up a bit to clear his head. As Street watched from above, Kerr continued down to the seabed and swam in circles through the mushrooms, picking one up along the way. Less than a minute later they were heading back up.

An hour and a half later, the two men broke the surface. Holding a rope that dangled from the transom, Kerr dropped the regulator from his teeth and pushed his mask onto his forehead. Twenty faces crowded around the gunwale, looking down expectantly, Ong at their center.

"So? Any luck?" Ong asked.

"Well, we didn't have a lot of time down there. Wasn't much to see," Kerr replied. He watched Ong's face fall for a few seconds before producing what he had plucked from the bottom: a bowl.

Ong's face cracked into a wide smile. He took the bowl reverently from Kerr's hand.

"Very good. Very good. Congratulations." His eyes widened slightly as he examined the find, as if its lines might reveal the answers to all the questions that must have been crowding his mind. How big was the ship it came from? How much was intact? Kerr could offer him only the bowl itself; time had been too short to do any exploration. As he and Street peeled off their wet suits, the Vietnamese committee members passed the piece excitedly from one to the other. With their history all but erased by two thousand years of conflict and occupation, seeing the bowl emerge from the depths unaffected by the tumult was a moving experience. Vietnam had been free when the piece had been created, and was free again now. China remained a threatening presence to the north, but the pot symbolized their independence. Even on its own the bowl was a treasure—the prospect that there might be thousands of such pieces on the seabed was intoxicating. Only the skipper and his seamen hung back. They knew they would no longer profit from the wreck. Besides, in their superstitious eyes, there was a difference between being given what landed in their nets and taking it for yourself from the seabed.

Ong took pictures of this first find in everyone's hands and had photos taken of himself, grinning broadly as he stood between the two divers. The museum experts on board confirmed to Ong that the bowl was of the same style and date as the pieces the fishermen had been selling. The boat headed for the mainland, everyone on board feeling jubilant. As soon as his mobile phone was within range, Ong began a steady stream of calls. The wreck had been located. With no time to waste he began to swing the operation into action, arranging meetings with ship-yards, dive contractors, and government officials.

OF ALL THE tasks facing Ong, the most critical was to recruit an arche-ologist to direct the excavation. He'd already mentioned Mensun's name to the Vietnamese Ministry of Culture, which was extremely keen: Men-sun enjoyed a high profile in Southeast Asia as a result of the *Diana* case and his work on the *Nassau*. However, there was a problem. Those in the top posts at the world's archeological faculties, including Oxford's Insti-tute of Archaeology, had reached a consensus that ancient shipwreck sites should be left untouched until more advanced techniques became available. Only in occasional situations, such as when a wreck was in par-ticular danger from coastal development or changing environmental cir-cumstances, could an excavation be justified. Even then, it should be commenced only if sufficient resources were found to see a full excava-tion through from pre-disturbance survey to exhibition. This might take ten years, but at least no information would be wasted. Excavations that took only a few months and left much of the hull on the seabed—as with the *Nassau*—were seen as unethical and betraying the cause of archeology.

Mensun saw it differently. He had reacted strongly to the looting of the Etruscan ship on Giglio; his whole perspective had shifted. He threw away the rule book developed by George Bass that had prescribed the different drawings, photographs, and measurements that should be taken from an artifact in situ. Valuables could not be left exposed on the seabed overnight, he decided. While archeological sites on land could be fenced off and guarded, protecting an underwater site was difficult and expensive. Sometimes it worked. Mensun remembered with pleasure the German looters who surfaced early one morning above one of his excavation sites to find themselves looking up into the gun barrel of an Italian *caribinieri*. But even having watchmen on board a surface vessel could offer no guarantee, as he later discovered when determined looters crept into a site on underwater scooters from a hidden bay.

Before, Mensun had joined with those who believed that only perfection was good enough; now he felt that information had to be saved from the seabed while there was still information to be had. Archeologists weren't the only people exploring the seabed, and sites were disappearing. Soon after Giglio, the Italian archeological authorities had asked Mensun to look at what they suspected was a medieval wreck, but Mensun turned down the request, feeling that he did not have time to see the project through. Three years later the wreck was gone, lost to looters, and two decades on he was still cursing himself for not having at least taken a look. Working faster was preferable to losing the sites entirely, he had decided. The maritime archeologists who were regularly leading excavations around the world could be counted on the fingers of two hands, but the number of looters, souvenir-seekers, and well-equipped treasure-hunters was in the high hundreds. In an ideal world Mensun would spend ten years excavating a wreck, but, after witnessing the destruction caused by thieves, he concluded that with a battle being fought desperate measures had to be taken.

When Ong called Mensun and told him that he'd found a wreck carrying Vietnamese ceramics near Hoi An, Mensun immediately thought of the mandarin he had seen in the Oxford laboratory. Had the statuette come from the Hoi An site? The thought of a cargo full of such rare and expressive painted pottery made his heart quicken. Here was almost virgin academic territory. If indeed this was the source of the mandarin and all the other Vietnamese wares that had been appearing on the market, then it would be the most important wreck Mensun had ever investigated, the most remarkable Asian wreck ever to have been found, and perhaps even one of the five most significant wrecks ever. It wouldn't, he knew, remain untouched for long. Someone else could get there first.

But there was a problem. Mensun wanted to rescue shipwrecks from what he saw as inevitable looting, but Ong was suggesting more than just a rescue excavation. He wanted to sell much of the cargo with an eye to making a profit, and that changed everything.

The UNESCO convention had set out strict ethical guidelines for underwater archeologists: No excavation should seek to make money from its finds, and artifacts collected from a shipwreck should always remain together. Getting involved in a project in which artifacts were to be sold and dispersed contravened both of these cardinal rules. The underwater archeological establishment had become militant in its fight against treasure-hunters. Any archeologist defying that establishment would be risking his career. Were Mensun to get involved, any research papers he submitted would be rejected by the academic journals and he would stop being invited to conferences. The very importance of the wreck steeply raised the stakes, for the institutional condemnation for destroying a site of such value would be enormous.

Mensun hesitated, needing time to consider, but Ong began to press home his perspective, selling the idea of commercial archeology. He explained that Vietnam would never be able to excavate the wreck near Hoi An for purely archeological reasons; the operation was simply too expensive. Why not collect all possible information from the artifacts, publish it for academics to use as reference, and let some of the pottery be sold to cover the costs? He pointed out that in the world of science, many researchers were coming to terms with commercial realities and teaming up with business; there was no reason why underwater archeology should not do so as well. If nothing was done to salvage the Hoi An wreck, the fishermen would soon return and everything would be smashed or sold on the black market. Selling some of the pottery from the excavation might not be perfect archeology, but it was better than losing the wreck entirely. Besides, argued Ong, rescue excavations were established practice on land. If a developer discovered Roman remains while digging foundations for a new skyscraper, archeologists would undertake a rapid excavation before the concrete was poured. The work was usually funded by the developer, who generally imposed tight time limits and financial pressure, but these were seen as necessary evils.

Anxious to avoid the prospect of a commercial project, Mensun suggested to Ong that given the importance of this cargo it might be possible to interest a foundation in sponsoring the work; that way, nothing would have to be sold or dispersed. Ong answered with a simple question. The Hoi An wreck might be carrying as many as 100,000 pieces. Where could such a quantity of pottery be stored or displayed? Mensun was forced to agree. There would be far too much to display in a museum and, having been a cargo manufactured in order to be sold, many of the Hoi An items would be duplicates. In Italy he had seen enormous storerooms in the basements of museums where beautiful relics gathered dust, appreciated by no one other than the curator. Museums were already in crisis, their storerooms and display halls filled to overflowing, their directors forbidden to sell off anything from their collections.

The more Mensun thought about it, the more Ong's viewpoint made sense. All over the world wrecks were fast disappearing at the hands of trawlers, dredgers, and cable-layers as well as treasure-hunters, while archeologists stood by helplessly, lacking the funds to work. The UNESCO convention was intended to stop trade, but Mensun had always thought the notion highly unrealistic. The market could not be switched off like a lightbulb. Artifacts from shipwrecks were intrinsically appealing, and therefore valuable. Were the trade banned it would be driven underground, and there would be even less chance of controlling it. The answer lay in regulation, not prohibition. Like Margaret Rule, the English archeologist behind the *Mary Rose* excavation, Mensun realized that if an artifact had been properly documented, there was no reason that it shouldn't be sold. Even a developed nation like Britain was having a hard time managing its underwater heritage, and non-archeologists were feeling increasingly excluded. That was a dangerous state of affairs, for generally it was the fishermen, salvagers, or amateur divers who found shipwrecks, not archeologists. Mensun knew one salvage diver who had come across an eighteenth-century man-of-war in beautifully preserved condition, freshly exposed by the shifting banks of the Goodwin Sands off Kent. Fearing that if he were to report the find it would be taken from him with little or no compensation, the man kept quiet and pulled the bronze cannon from their still-intact gun carriages and smelted them for their scrap metal value, never telling anyone of the wreck's location. There was nothing to stop the same thing happening again in other places, and good reason to suppose that it did.

Mensun began to consider some of the finer details. The cost of the work that followed actual excavation could be high; in fact, conservation, analysis, and publication frequently amounted to two-thirds of a project's total budget. He needed to know that Ong was not going to use MARE's involvement as a rubber stamp to get past the Vietnamese government, only to walk away without paying for the post-excavation expenses. Ong assured him that the Vietnamese Ministry of Culture would not grant an export permit for the ceramics until Mensun's own excavation report was submitted. He explained once again his vision that Mensun's academic publication would be available at the auction. Auction houses were now keen to prove the provenance of the items they sold, while the ship's story, the historical background, and scientific analysis helped in marketing.

The archeological politics aside, Mensun worried about the practicalities of excavating a wreck that was deeper than any archeological excavation had ever been performed. Yes, there had been plenty of salvages at greater depths, but because of the huge expense of working at such pressure they were smash-and-grab affairs. He had to be sure that his reputation would survive a project like the Hoi An wreck. Working at 230 feet underwater, he wasn't sure he would be able to practice what he had preached during the *Diana* trial.

Ever since his work with George Bass, Mensun believed archeology to be about expanding knowledge and gaining new information. Turning the project down would not only be refusing the opportunity of a lifetime, it would also be irresponsible. He had seen the situation on the coastlines of the developing world. Those behind the UNESCO convention had, he felt, closed their eyes to the realities there. The destruction was happening; putting your head in the sand and waiting for a perfect world was not going to help. Contemporary concerns and politics should not allow him to let an entire unique cargo be destroyed. Mensun decided to accept Ong's offer.

Part Two

CHAPTER 7

Dockyard Drama

I GOT A CALL from Mensun one day in early 1998. I'd known him for six years by this point, long enough for me to tell from the tone of his voice that something exciting had come up. Every summer during my four-year zoology degree course I'd worked as a volunteer on MARE projects in the Mediterranean—initially on a medieval wreck on Zakynthos, then on a survey of the deep water surrounding the Italian prison island of Gorgona, where ancient anchors hinted at the presence of a wreck. During the search we stayed in a prison guard's house and ate in a mess served by the prisoners (one of whom would pass notes to me, protesting his innocence and pleading for help in escaping). By the time college was over I knew that the sea was where I wanted to work, and I went to lead a coral-reef survey expedition on a small coral atoll in Belize, Central America. Four months later, just before Christmas, as I was about to sign on as a project manager for the organization, I had received a fax from Mensun.

The smudged print stated that a big project in Uruguay was just starting. The excavation of HMS *Agamemnon*—Nelson's first command that
sank in 1809—was the focus, but we'd also be doing a filmed investigation of the World War II *panzerschiff Graf Spee*. I wasn't hard to persuade,
and within a week I flew down to join the team. The divers, draftsmen,
and administrators were all volunteers, using their vacations or their
pension plans to enable them to come out and work, but I soon realized
that I was more serious about it. I wanted to make shipwrecks my life
and my living so I stayed in Uruguay until I started getting paid. It was
with Mensun and MARE that I now saw my future, and I began doing
everything I could to make myself indispensible. I started bringing new
technology into our operation—digitizing site plans and imaging, building project Web sites and databases.

After the Uruguay project I had gone back to London, but for two
months there was no work with MARE and I began feeling landlocked.
Finally Mensun called. He spoke in one long stream, describing in detail
the Hoi An wreck project and waxing lyrical about junks, ceramics, and
hard-hat diving. I felt the grip of the land loosen. Mensun hadn't been
able to tell me anything about the endeavor before because nothing had
been settled, but now it had, he explained. All I needed to do was to
meet his partner Ong Soo Hin and convince him that I could help make
the archeology more streamlined and efficient, and I'd have the job.

THREE WEEKS LATER I was in a cab leaving Singapore's ultramodern
airport behind me, staring in wonder at immaculate grass verges that
bordered chewing gum—free sidewalks, watching lines of citizens waiting patiently at pedestrian stoplights to cross car-free roads. To the accompaniment of the taxi's speed alarm, a merciless "ding, ding, ding,"
the antiseptic cleanliness of the city center slowly dissolved into more

fragrant and dusty suburbs. The driver had evidently grown deaf to the alarm's piercing chime, and eventually I had to ask him to slow down. Able to think once more, I settled into watching the world roll by. I'd been hired as a "finds manager," and would be under Mensun's supervision but with my salary paid by Ong—a comfortable enough arrangement. My one gripe was that I would be working on the surface, not the seabed. Still, friends within MARE who'd been on the *Nassau* in the same position told me they'd been able to spend some time on the wreck, so I held out hope.

We drove on west toward the dockyards, a silent electric monorail train snaking along the road beside us. In the city center its track slid invisibly beneath the gleaming skyscrapers; out here, it rose disdainfully above the bustling streets and the multicolored plastic quilt of food stalls in the market. Those living in the suburbs had to climb above the happy chaos every day to connect with their city's aspirational transport system. Finally I glimpsed water. The brown fingers of the sea that penetrated the docklands were oily and placid, their banks crowded with steel cranes like clumps of giant bulrushes. Eventually the cab slowed and turned in through some rusted gates. We'd arrived at the Pandan Road dockyards. I got out, paid the driver, and began to pick my way through tangled piles of iron and puddles of oil, looking for the right quay. At last I found it and was greeted by Dilip Tan, the operations manager.

A fresh-faced Singaporean-Chinese who spoke in short, stabbing sentences, Dilip showed me where to stow my gear, then took me around the barge that would be home for the next few months. At 180 feet in length, *Abex* was the biggest platform I'd ever worked off, and a far cry from the inflatables and onshore bases to which I was accustomed. Stacked shipping containers at the stern had been converted into air-conditioned offices, bunk rooms, washrooms, and a mess hall. The containers formed a square enclosure two stories high; a tarpaulin roof was stretched over it to provide shelter. Forward of the living area was the business end of the barge. The foredeck was filled by huge torpedoes of gas, three more containers carrying decompression chambers, and the dive-control room, where a myriad of valves and gauges covered the control panel. Everything had been carefully painted—the deck bright green, containers white, and all the equipment color-coded: electrical in red, hydraulic in blue, and pneumatic in yellow.

Dilip, who had met Ong while working as a diver on the *Nassau*, was in charge of "mobilizing" the barge, but that evening he grumbled that although everything was ready, we couldn't leave. Over the previous few days he'd been having trouble in a number of areas: crew members' paperwork, gas hire, and seaworthiness certificates. It felt, he said, as if someone kept throwing wrenches in the works. The next afternoon I was talking to him on deck when suddenly he broke off, leaped onto the quayside, and began marching toward a burly man who was approaching *Abex*. Dilip blocked his path, and there was an altercation. I wasn't close enough to hear what was said, but the man eventually shrugged, turned around, and left. When I asked Dilip what had happened, he glowered. That was a guy named Mike Hatcher, he said, who he had been told to keep away from the barge at all costs. He hadn't realized the man would actually show up, but now that he had, Dilip reckoned he knew the source of his recent troubles.

The paperwork was finally cleared and the next day a tug pulled *Abex* from the crane-bristled Pandan Road dockyards, through the Sembilan Straits, away from Singapore and into the South China Sea. The coastline of Malaysia slipped away as we crossed the Gulf of Thailand, the barge silent apart from the thumping of waves on the hull and the distant rumbling of the tug far ahead. I was the only member of MARE's team on the tow, finalizing the artifact databases and logging procedures as we sailed. My shipmates were the Indonesian deck crew and the commercial divers who would be carrying out the underwater work.

All the diving I had ever done had been with scuba. I'd always assumed this was the most evolved form of diving, persuaded by the popularity of Cousteau's equipment and the romance of the films of his adventures with the other "menfish." But talking to the commercial divers in the mess on board *Abex,* I began to feel otherwise; I was in-trigued by this parallel universe of hard hats in which diving was so dif-ferent. The commercial diver could communicate freely with others underwater and with the surface rather than operate, scuba-style, in a world of mute signaling and scrawled messages on slates. He moved as fast as the water allowed and could breathe as hard as he liked, rather than preserving the careful calm that rules dives with a bottle and lim-ited gas. He walked along the seabed, rather than floating above it. All in all, diving with a surface supply seemed to be less of a touristic visit into another element, and more of a working visa—if not quite a resi-dent's permit. Technical scuba divers might sometimes visit deep sites such as the SS *Britannic* at 360 feet, I was reminded, but the conditions had to be perfect and they could only stay long enough to take some photos, swim across the deck, or peek inside. One of the deepest scuba dives ever was made by an Englishman named John Bennett in 2001, who mobilized an enormous logistical effort to descend to a depth of one thousand feet. The dive took him nine and a half hours, but he spent less than a minute at the bottom. Commercial divers not only visited Bennett's record depth routinely, but worked there too.

Whenever I'd done decompression dives—those requiring stops on ascent—it had meant spending long periods hanging off a rope in mid-water. To come up without stopping, I'd always understood, would be to invite the bends. Not so, snorted Jack Ng, a tall Singaporean-Chinese diver with a Maori-style nose ring and tattoos covering most of his body. The rules of the bends could themselves be bent. On this job he'd be doing a few stops in the water, but not nearly his full decompression schedule. Instead he would surface and take off his gear, gas already fizzing out of his muscles and starting to form the bubbles that could kill him, have a bucket of fresh water thrown over him, and then climb into a chamber on deck. As soon as the hatch was shut the chamber would be pressurized, forcing the gas back into his tissues. Only then would Jack restart his slow decompression, following the dive tables. As long as he didn't spend more than two minutes at surface pressure getting into the chamber, he would be fine. Breathing gas supplied from the surface didn't change the need for decompression, but having chambers on deck meant that it could be done in safety and comfort rather than drifting in mid-water. All the divers would be routinely doing their decompression on deck—which explained, I realized, the three pressure chambers that were mounted near the front of the barge.

Dilip kept the divers busy during the tow, making them strip down every piece of equipment, clean it, and put it back together. Compared to the grizzled and tattooed antipodean divers, the forty-year-old at first seemed too young to be in charge, but he had in fact worked in the offshore-diving industry for eighteen years and his actions broadcast a physical confidence born of a dangerous line of work. After the monotony of the endless pipeline surveys and wellhead inspections of standard offshore contracts, he had found working on the *Nassau* fascinating. Initially hired as a simple diver, he had stayed on and became the supervisor of all diving on the project, taking over when Nigel Kerr got busy with other aspects of the job. Dilip had worked on shipwrecks before, at one time for the Japanese Yakuza, scouring a World War II wreck in search of gold. Unfortunately, their information had been misleading, and when they blew apart what they had taken to be the gold storage hold, they released instead an enormous slick of oil.

Ong had tried to get Kerr to run the operations on board *Abex* during the excavation, but was unable to match the terms of a lucrative security contract that the ex-Legionnaire had been offered in the Algerian desert. Ong had turned to Dilip instead, hiring him as the operations manager. Both he and Dilip were of Chinese descent, and there was an implicit trust between them. Ong encouraged Dilip to set up his own dive-services company, and then to take on the contracts for their shipwreck work. Dilip did so, and because his fortunes were now connected to those of his boss he pulled out all the stops, begging favors from old industry friends to get gases at cheap rates and hire equipment at a discount. This wasn't the way he was used to doing his job. With oil companies, the money had always been available to get the best equipment. Ong, on the other hand, seemed to be constantly looking for a cheaper way. He questioned everything. Some of the time this was fine, but Dilip knew about diving and the dangers of working at sea: Some economies were false economies. Still, Ong was the boss. He explained that the funding of salvage was different from that of oil-sector work. Given the uncertain returns, costs had to be watched carefully. Dilip's requests to use a more substantial vessel and more expensive diving system were turned down. The crew would have to take the risk of using the cheapest methods and equipment possible, and hope their luck would hold.

CHAPTER 8

Lost and Found

TWO WEEKS LATER I watched enviously as two divers stepped onto the "stage," the open elevator that would take them to the seabed. They seemed dwarfed by their bulbous yellow headgear, from which sprouted multicolored cables and hoses carrying breathing gas (in this case a helium/oxygen mix) and power for their headlight and camera, communications, and video data. The cables were bound together in a single umbilical that ran across deck and into the back of a control panel that regulated the gases, showed camera images, and housed the diver telecom.

The system was both beautifully sophisticated and archaic at the same time. The basic technique of feeding a helmet with air from the surface had been in use since the seventeenth century. The original helmets were open around the neck, the water kept at bay simply by keeping enough air pumped inside. As long as the air wasn't spilled, the system worked fine. Later helmets were sealed to a leather diving suit and allowed the diver more flexibility in his work, as long as he had enough air

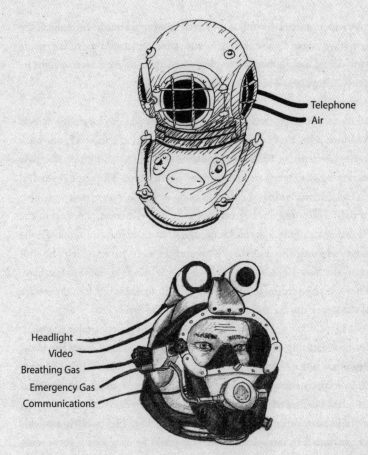

Telephone
Air

Headlight
Video
Breathing Gas
Emergency Gas
Communications

inside for his limbs to move freely. Simple air-filled leather tubes running up to the surface enabled rudimentary communication with the support staff. Later, however, Siebe-Gorman invented a telephone that revolutionized diving work. Supervisors were able to talk to one diver at a time, or both; it also allowed one diver to speak to the other. Not until the 1970s were the heavy bronze helmets finally abandoned. The replacement currently favored was the Kirby-Morgan Superlite 17b, a tough carbon-fiber shell that manages to carry the menace of a *Star Wars* storm trooper's mask, despite being bright yellow.

A winch started up and, as two deck hands paid out the umbilical, the dive stage began to descend. A minute later, all I could see of the divers were occasional flashes in the blue when their helmets were caught by shafts of sunlight. Then they were gone.

BUT ALL WAS NOT WELL. The divers were unable to locate the wreck. A month prior to *Abex*'s departure from Singapore, Dilip had gone back to the site with an ROV to reconfirm its location and dropped a new marker buoy to guide our positioning of the barge. However, on arrival we had discovered that the marker buoy was gone, either stolen or sunk. Initially Dilip had not been worried; he had the coordinates from the Global Positioning System. Using signals from an array of satellites, the hand-held unit should have got around the need for anything else. But when the divers reached the seabed they could find nothing but mud. They searched to the limit of their umbilical cables but found no trace of the wreck during their entire forty-minute dive.

When Dilip double-checked the GPS reading, he found it was wandering erratically. He brought out the backup unit, and it did the same. Rumors—not confirmed until years later—were that in times of conflict or international tension, GPS units became much less accurate. GPS had been developed by the American military, which used it to determine positions accurate to within three feet. The publicly available data included an intentional error that could be increased if there was a fear that hostile forces might be about to use the system against them. Dilip had no way of knowing that on the very day he was searching for the wreck, on the other side of the world the United Nations Special Commission (UNSCOM), established to control Iraq's nuclear ambitions, had learned that an Iraqi delegation was in Bucharest meeting arms dealers who were selling GPS-based missile guidance systems. The United States was in effect jamming the system with a massive error, rendering the missiles inaccurate. As a result, Dilip was unable to tell if we were right on top of the wreck or five hundred feet away from it. Eventually he gave up on the GPS and began searching with the ROV.

Within a day the ROV broke down, just as it had during Kerr's first search.

Dilip cursed. He had wanted to mark the site with an underwater beacon, a "pinger" that would lead unerringly to its location, but Ong had considered its $50,000 cost too high. Dilip resorted to moving the barge. Though *Abex* was reliant on the tug when it came to leaving the site, she was able to move around a limited area on her own by pulling on her anchors. A cable stretched from each corner of the barge, anchored to the seabed hundreds of feet away. By winching in on some cables and out on others, the barge could haul itself around like a spider on its web. Guessing that the wreck must be close, Dilip began to search by tracking back and forth with the barge on its anchors, dangling a drop camera from the bow. It was a slow process, not helped by the currents and winds that worked against the hull, straining the winches and causing the anchors to slip. Eventually he called the support tug *Atsa* to help push the barge, but this only added to his woes. Instead of nosing gently up to the side of *Abex* the tug rammed her, puncturing the hull just above the water line. Dilip was furious at the tug captain's incompetence, but knew that he could hardly expect more for only $750 per day. A proper offshore tug and crew would cost at least four times that.

Ong's mood darkened with every day the wreck evaded discovery. In the evening he would emerge from the barge office that he shared as a cabin with Dilip and begin a punishing series of exercises on deck. As he progressed through his squats, curls, and push-ups, he would remove his shirt, revealing a compact frame bulging with outsized muscles. The Indonesians and Vietnamese averted their eyes, unsure of how to react to their boss's public workout.

After five days, Hoang Van Loc, VISAL's operational director, put a friendly arm around his partner's shoulder. "Things aren't always easy at sea," he said. "Especially this sea. You have a heavy spirit. Perhaps you are weighing down the project with your worries. You should spend a few days ashore. See if things change for the better." Ong decided to take Loc's advice. He would try anything that might turn his luck around. He joined the small wooden supply sampan *Song Hoi* 3 for the three-hour trip back to the mainland.

Dilip located the wreck with the drop camera the next day. Ong was notified and raced back out to the site while the divers made preparations for another descent. Two hours later I stood behind Ong and Mensun in the dive-control container, watching the descent on the monitors. We caught occasional glimpses of the stage's frame or a gloved hand while in the background the water darkened. Then Dilip gave a signal and the winch slowed. A few moments later the seabed appeared.

Mud stretched as far as the divers' headlamps could illuminate. The seabed looked like the moon; the only visible features were wormcasts. There was no sign of a wreck. The two divers struck out in different directions, and within minutes one came across a deep gouge in the mud. There was a murmur of conversation among those gathered. "You think it's from a dragging anchor, Diver One?" Dilip asked into his headpiece. There was a pause, and the image jerked upward and began to move. Another gouge appeared at the edge of his headlamp's range.

"They run parallel. They're not anchor tracks," Mensun said. "They've got to be from the fishermen's rakes." No one replied. Just beyond the second gouge the diver reported that he could feel things underfoot. A few minutes later he stopped. Specks of illuminated sediment swirled around his helmet, like sparks from a bonfire, then the image steadied. His hand reached down and mud clouded up from the seabed. Two seconds later a fragment of ceramic appeared, slowly turning in the diver's fingers.

Mensun and I looked at each other. The breaks at the edges of the fragment were clean, devoid of marine growth. Their freshness suggested that the fishermen had visited recently. Suddenly we feared that all we'd find were fragments scattered across the seabed, the site's coherence smashed into pieces like the remains of the dish the diver now held in his hand.

"Let's get a grid in place before he does any digging," said Mensun, leaning over Dilip's shoulder. Ong shifted uncomfortably. He badly wanted to know if there was something left intact beneath the mud, but he bit his tongue.

A square steel grid with sides measuring two meters, or about six feet, was lowered to the two divers, who adjusted its legs to lay it level over the area. It was the first of the grids that would later cover the seabed. On most excavations the site would be mapped by measuring every find or feature to three fixed points on the seabed, enabling its position to be plotted. Here, because of the nature of the wreck, Mensun and Ong had struck a compromise. Each grid would be cleared of mud and the fragments lifted into a basket for storage according to which grid they came from. Then, after a sketch had been made of what the grid contained, that layer of pottery would be removed and brought to the surface. Each piece would then be given an individual number that referred to the layer and grid in which it had been found, and on what date.

The divers began work in one corner of the grid, placing the smashed fragments into a plastic basket. On deck, no one spoke as we waited to learn if the damage continued beneath the surface layer. The diver's gloved hand swept eddies of mud into the water, then paused, letting the current carry them away. From among the swirling clouds, rows of short white lines began to appear in the brown. As the diver moved his head closer, it became obvious that they were rims of dishes. There was a collective sigh of relief. Ong was smiling, and Mensun's face was taut with excitement. "Bingo! Look at that! Stacked on their sides just as they were packed!"

As the divers continued to clear the debris, a wooden wall was revealed at the end of one of the rows. Mensun was close to the screen again, anxious to know more. Dilip relayed his requests for information on the thickness of the wood and the direction of the grain, and Mensun listened to the divers' replies, nodding and taking notes. Ong folded his arms, a frown on his face. He was evidently anxious to get some pottery up, but the dive was already nearing its end. Mensun was oblivious.

"Okay, probably structural. Now tell me, do you see any evidence of woodworm on there?"

"You've got about five minutes before I start bringing the guys back up," Dilip said, glancing at Ong.

"Okay. Just get him to hold the camera there," Mensun replied, and started making a sketch of the dish rims and the piece of wood in relation to the edge of the grid. Ong began shifting from foot to foot.

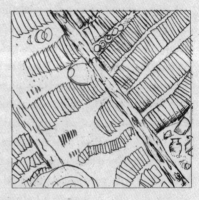

When Mensun had finally completed his sketch, he asked the diver to lift out one of the dishes. They were very tightly packed, but after some gentle wiggling the diver managed to pry one free. He fanned the mud from its surface, then held it up for his helmet camera. The screen flared bright as the lamp caught the white ceramic, then the lens adjusted and found its focus. Slowly, from the mud of five centuries, an irate-looking lion appeared, smoke billowing from his mouth and clouding the background of the dish. The design leapt off the monitor, a vision of flowing lines in the middle of the knobs and valves of the control panel. The decoration was astonishingly crisp, and the sight of such an object being lifted from the mud, having lain untouched for hundreds of years, was eerie. I was used to artifacts being all but unrecognizable after their long contact with the salt water, half rotted away and covered in marine growth. This glazed ceramic looked as if it had just come from the kiln.

"Whoa. Look at him!" Mensun said, half to himself.

Ong nodded. "Mmm. Very nice. Ve-ry nice. Isn't it? The quality not too bad, mmm?"

Mensun nodded. "It's wonderful. Look at the artwork, Ong. It's beautiful," he said, not taking his eyes off the dish as the diver turned it in the light of his lamp.

THE NEXT DAY, the *Abex*'s entire complement of forty turned out on the foredeck, where an altar had been set up. Flowers brought from the mainland nestled among fruit, glasses of rice wine, burning candles, smoking incense sticks, and a plucked, uncooked chicken. A priest struck a gong, its tone sounding out into the mist that hung over the oily-calm sea. Loc, VISAL's director, stood beside the altar dressed in a long red silk gown and a matching cap.

"It is our tradition to beg for mercy and forgiveness from the Emperor of the Sea," he explained to the congregation. "We have entered his territory and taken things away. To appease him and to be permitted to continue, we must pray." When Loc finished, Ivo Vasiljev, now on board as the expedition's official translator, recounted in Indonesian what had been said for the deck crew. Ivo, by far the oldest of those on board, had only been on short ferry crossings before; his eyes gleamed with excitement at his situation.

Ong and Mensun stood immediately before the altar, ringed by the assembled team. Both had dressed for the occasion. Ong wore sunglasses and a brand-new pair of white expedition overalls. Mensun was in a pale yellow shirt and his best white jeans. Only the frayed piece of string around his neck that he wore as a camera strap betrayed his usual style. His hair was even brushed.

Loc began to speak in gentle, lilting Vietnamese, bowing rhythmically to the priest's gong. On cue, Mensun and Ong both put their palms together in prayer. While Mensun's face was a study in solemn contemplation, Ong's sunglasses and half-formed smile implied skepticism. Behind them the Vietnamese museum staff and artists had closed their eyes, their lips working as they followed the service in silence. The Indonesian deck hands and galley crew simply bowed their heads in respect. In loose clusters at the edges stood the dive crews, shifting on their feet and peering over shoulders like tourists observing a cathedral service.

At last Loc's chant finished and Mensun and Ong retreated, walking backward from the altar before turning away. The crowd dispersed. Soon the serene silence was replaced by the chugging of generators and dry roars of escaping gas.

JACK NG, the tattooed Singaporean-Chinese diver, was next in the water. His task was to investigate the extent of the site. Equipped with one hundred feet of umbilical, he walked straight out from the stage due east to the limit of his cable, then back to the stage and out west, then north. Wherever he went, he walked over pottery. The site was enormous, far larger than anyone had expected. The mound alone stretched more than ninety feet from northeast to southwest and almost thirty feet wide, and more ceramics were buried in the spill outside. Better still, it seemed that the fishermen's rakes had not been able to penetrate the hull walls, thus everything inside had been protected.

Satisfied that things were on track at last, Ong left the barge to report back to his investors. On Sunday evening the chef announced that he would prepare a barbecue on the stern to celebrate the long-awaited start of operations. That night I found Dilip and he accepted a beer, his eyes smiling for the first time since our arrival ten days before. He was relieved that the first ceramics were now on deck. Things seemed to be going right at last, and we talked cheerfully about the plans for the next day. Meat was sizzling on the grill when the first spats of rain hit the fire. Though everybody laughed at the weather's timing, the party petered out soon after the eating was done.

CHAPTER 9

The Dragon Stirs

WHILE THE REST of us were in bed, Dilip remained alert, watching the weather build. The most recent forecast from the Weatherfax service had been unclear, the page speckled with a spray of dots that obscured the printout. It could simply have been a bad connection. Sometimes, however, stormy weather could produce enough static electricity to interfere with the satellite link. Since the report that had come through four hours earlier showed nothing more than a slight depression to the northeast, Dilip wasn't worried. Nonetheless, he had decided to remain awake to keep an eye on things.

As the night wore on the wind picked up steadily, and it was soon pushing insistently against the barge's superstructure. Normally an anchored ship hangs off a single point and is free to swing about, finding her own path of least resistance through oncoming weather. *Abex* had four anchors down to lock her in position over the wreck. She wasn't free to ride anything out.

The brakes on the bow anchor winches began to slip. The winches weren't the specification Dilip had initially requested, but he had found himself unable to justify large safety margins in the face of Ong's drive for economy. On paper, they were strong enough. Dilip was now forced to make a critical decision: Run for shelter, or sit tight. He could put the winches into gear to stop them from slipping. Once engaged, however, there would be no way of releasing them again until the weather had calmed and the tension was taken off the cables. If the weather got worse and he had to move, he would have to cut the cables to free the barge. That would not only be expensive—a single thousand-foot drum of steel cable cost at least $10,000—but time-consuming. The cables were too expensive to carry as spares. Cutting even one meant a week lost in getting a replacement, and even more time to get it spooled onto the winch drum. On the other hand, if this was the worst the storm had to offer, putting the winches in gear would let the barge ride through it, and the team would be ready to work as soon as the weather cleared.

The other option was to leave immediately. Dilip could get the tug to pick up the towline that was attached to the front of the barge, then loosen off all four anchor cables and get towed to shelter. Recovering the anchors and their cables from the seabed after the storm would take at least two days, probably three. Add another day for lining the barge up over the site, and that was as many as four days lost. Mindful of having lost five days looking for the wreck in the first place, Dilip decided to sit out the weather with the winches in gear and hope the conditions would calm.

They didn't. Both wind and waves continued to rise. Dilip began to worry about the punishment that the barge was taking. The swell was hitting her at an awkward angle and its impact caused the whole hull to vibrate as though struck like a gong. *Abex* had been built as an inshore barge, and this was not inshore weather. Her age didn't help. Dilip had been shocked when carrying out the hull-thickness tests required to gain permission to leave Singapore. He had had to replace the bow platings and the internal bulkheads, and when he did so the metal wilted away from the welding torch. The rust had been so deep that it had been hard to know where to stop the repairs.

Abex was carrying a heavy load, her deck only three feet above the sea in calm water. The entire barge flexed with every wave that passed beneath her, but not everything on deck could do the same. Because Ong had experienced a problem releasing funds during the mobilization, Dilip had been able to buy the diving gas only at the last minute. Instead of racks of gas in small vertical cylinders, it had been delivered in forty-five-foot-long torpedoes contained within a massive rigid frame. Some of the frame's spars were twisted already and though Dilip hadn't yet seen them, hairline cracks were developing in its welds.

Just before two in the morning, Dilip took a look at the stern anchor cables. They should have been slanting away from each corner at a taut, shallow angle into the water, but instead they looped loosely under the barge. The platform was moving backward. The bow winches hadn't slipped—they were still in gear. There was only one explanation: The bow anchors had to be dragging across the seabed. When he looked again not only were the stern cables slack, they had twisted around each other. This was bad news. When the barge passed the stern anchors, those cables would come tight and would drag the stern down or, worse, flip the vessel. Dilip ordered the crew to start winching in the stern anchor cables as fast as they could and slammed the general alarm with his fist.

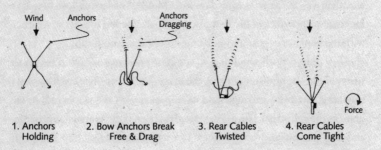

1. Anchors   2. Bow Anchors Break   3. Rear Cables   4. Rear Cables
   Holding   Free & Drag   Twisted   Come Tight

I WOKE, my head filled with the screaming siren. I scrambled out of my bunk, the ear-splitting noise crowding out any other thought or sound. Then an ominous vibration shook my stomach from up my legs.

Everyone in the cabin seemed to freeze for a moment and I caught Mensun's eye. He had felt it too.

I stepped from the container onto the grated metal walkway outside. Instantly I was flung sideways, soaked as sharp gusts of wind whipped around the barge. Grabbing a rail with both hands to steady myself, I climbed down the stairs, salt-water spray shooting through the superstructure stinging my face like handfuls of gravel. On deck the crews were clasped in miserable huddles around anything secure, clutching the straps of one another's life jackets, with water streaming around their ankles as it ran down the deck. The wind tore at loose edges and exposed corners, snatching objects from stowage points and flinging them through the air. A life ring worked free and took off into the blackness; a broom spun past a cowering crewman. Another wave smashed into the hull, and the knots of people tightened visibly.

I pushed toward the bow, away from the deck lights and the clusters of scared faces. In the blackness, the full force of the wind felt apocalyptic, the power supernatural. For a moment I was staring down the throat of a roaring dragon. Then the barge lurched as another wave hit, sending up a wall of spray and reverberations up my legs. As I looked back along the length of the deck reflected in the deck lights, the whole hull seemed to twist. The hairs on my neck tingled. I began working my way back aft as fast as I could.

"Stay here! Watch everyone and don't let them leave this area," Dilip shouted at me as I reached the muster area. He was in yellow storm gear, his eyes blazing with its reflection as he stood with legs braced against the swell. A VHF radio was held cocked next to his ear as he yelled. He glared at the bosun and some of the galley crew who were clutching the galley steps.

"I just found that lot 'round by the starboard lifeboat, making to abandon ship! Fucking idiots! Do that and they've not got a fucking chance. What do they think? They think that tug's going to find them in this? Pick them up? One was wearing three life jackets! One 'round the head, one on the arse, and one on the legs. He would float upside down! Shit!" His radio screeched, and he shouted back into it before disappearing into the darkness toward the bow.

A minute later he was back.

"Jesus Christ! The tug's run for shelter. Can you believe those cheap shit arseholes! They're supposed to be circling the frigging barge, but they've gone!" He gave another blast on the siren to bring everyone close enough to hear him against the wind.

"Listen up!" Dilip shouted through the lashing rain. "Our situation is not good! We're going to try to cut ourselves free from three of our anchors, and ride out this storm hanging on one." He paused, as though he was about to mention that the tug—our only means of escape—was nowhere to be seen, but thought better of it. "We'll drag, but we should be far enough from land. Everybody stand to my right and don't move!"

Ivo, still clutching a life ring, stood beside Dilip and, straining his voice, shouted out a translation in Vietnamese. Everybody moved to the port side, pressing themselves up against the walls of the offices. Dilip swung his flashlight up toward the winches, illuminating a crewman crouched a few feet astern of them. A dagger of blue flame flared in the darkness, then a shower of sparks spat onto the wet deck as the oxy-acetylene torch began to cut. There was howling silence, broken suddenly by a deep twang. The freed cable shot backward. Its molten end sizzled across the deck, snapped through the rollers, and disappeared. The barge shifted, seeming even more unsteady than before while we staggered as one to the starboard side, this time bracing ourselves against the galley. Again the glowing tail writhed as it whipped off, leaving a trail of steam through the water on deck.

The stern cables cut, the danger of getting dragged under was gone, but the barge began a lurching yaw. When the starboard bow cable went, the effect was instantaneous. The wind seemed to lessen and the sea to flatten. It was as though we'd entered the eye of the storm. The deck felt stable at last—not level, but safe. There were no more sickening crashes as waves caught us broadside. Eventually people began to retreat to their cabins to dry off and wait for daybreak.

At first light *Atsa* reappeared close by, bucking in the still-heavy seas. She needed to pick up our towing bridle—no mean feat in this weather. Her captain maneuvered as close as he dared. Dilip tied a messenger line—a smaller cable—to the towrope and began trying to throw it onto the tug's foredeck. It took almost two hours for the two vessels to get close enough for him to land the line on the tug's deck. When he finally succeeded, the tug's crew hauled the line and then the thick rope through the water and onto their deck, where it was secured to the towing hook.

All day we crashed through lead-gray seas as we were towed toward Da Nang, the nearest safe harbor given the wind direction. Each time the barge ran down the face of a wave the towrope slackened, its three-hundred-foot length disappearing into the sea. Plowing into the next wall of water, *Abex* recoiled like a reluctant dog on a leash and the massive hawser thumped taut. If it snapped the recoil could kill, so the entire bow of the barge was off-limits. But the towline held. *Abex* made it to shelter.

FOR THE NEXT two months it was as though nothing about the Hoi An expedition could go right. *Atsa* was a towing tug, and therefore lacked the winch equipment necessary to recover our anchors. The Vietnamese tug that was chartered to do the work collided with another vessel while leaving Da Nang: she survived, only to suffer an engine-room explosion as soon as she got within view of *Abex*. Her crew soldiered on with the one remaining engine, but took twice as long to do the job. Replacement anchor cables took ten days to arrive. New brake linings were fitted to the winches and proved faulty. Twice we were back on top of the wreck for less than two days when the sea reared up with no warning, forcing a rapid retreat.

The bright colors of the equipment on deck became freckled with rust, slowly developing into an orangey-brown rash. The concrete lining of the hull storage tanks flaked off and rust fell in the water supply, so that at every wash our clothes darkened too. Gradually the color-coded overalls that had been issued to the different crews—white for archeological, orange for deck hands, and blue for divers—began to turn a uniform rusty brown.

Even when the weather did give us a break, progress was painfully slow. For every forty minutes a diver spent on the seabed he had to decompress for two hours in the water, then another hour in the chamber on deck. Even with three chambers on board, we were achieving only two or three man-hours on the seabed each day.

Dilip was determined to make the job a success, but he was getting more and more frustrated by what he felt was a lack of suitable equipment. In his eighteen years offshore, he had clocked more than three years' worth of time underwater and learned some lessons. Working in the deep sea, the one thing you didn't do was give up. You were sent down to do a job and you did it, whatever the conditions. His attitude had made Dilip one of Asia's most respected divers, and subsequently a supervisor in high demand. However, success meant that his experience of working with budget operations was in the distant past. To his mind, decompressing the men after each dive was a waste of time. Each dive wasted three hours in decompression. The only way of pulling off the job was to use a saturation system. That way the divers would have unlimited time on the seabed. Their body tissues still filled with gas, but by living at pressure they avoided the need for decompressing at the end of every dive. They could spend eight, twelve, fourteen hours on the seabed and once their shift on the seabed was finished, the men just entered the dive bell and closed the hatch. The bell was then brought back to the surface and mated onto the saturation system. Once the clamps were secure and the airlocks pressurized, the hatches between the bell and the saturation system were opened up and the divers climbed into their chambers. There they ate and slept, all still at the same pressure they were at on the seabed. They still had to decompress eventually, and that might take as long as six days, depending on the depth, but at least the process was deferred until the very end of the job, even if that was a whole month later.

But for all the advantages of using a saturation system, once again Ong had declared it too expensive. Using the wrong equipment made Dilip feel he had one hand tied behind his back, but he blamed Hatcher. It seemed to Dilip that Ong clung to the idea that if the treasure-hunter could pull off jobs like the *Geldermalsen* with minimal tools, so could he.

The last straw came in mid-June. Once more we had repositioned ourselves above the site. The air was calm and the sun shining. It was the break in the weather we had been waiting for, but I found Dilip hunched on a bollard at the front of the barge, a cigarette between his fingers. The contours of his face had sharpened and he had lost the boy-ish look I had noticed when I'd first come on board. His shoulders and arms were taut, veins bulging. I told him it felt like a perfect day.

"No, la. No diving. Cannot. You see this swell?"

There was a long, low swell running, but the barge was sliding over the glassy rollers without a sound. The motion was so smooth it was barely noticeable.

"If we'd had the system I wanted, then no problem. Like you say, a perfect day. But look. Look at the stage."

I followed his finger to the starboard bow, where the diving stage stood lashed to the edge of the deck.

"Tell me, how far between the deck and the water?" Dilip asked. He pulled his lips back in a bitter smile, revealing a nicotine stain between his front teeth.

I watched the bow rise as a wave rolled beneath us. It carried on up until the deck was ten or fifteen feet above the next trough, then began to slide back down. When the bow met the next roller, the water was no more than two feet beneath deck level. I stared, still not fully under-standing the problem.

Dilip shook his head, exasperated. "Decompression diving, right? These guys aren't coming up in a bell." They were out in the open, and

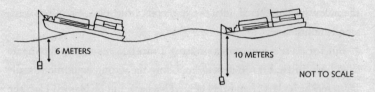

6 METERS

10 METERS

NOT TO SCALE

the stage dangled from the bow, he explained. Though they did most of
their decompression in the deck chambers, the divers still had to wait in
the water for two hours. Their last stop took half an hour and was supposed to be at a depth of thirty feet. That was the problem, said Dilip,
for how was he supposed to keep their depth constant? Sitting on the
stage, one moment they would be under thirty feet of water, the next
under twenty, maybe seventeen. That, he continued, was a sure way to
get the bends.

Now I understood. Using a smaller boat as a dive platform, you could
endure a fair-sized swell because your rise and fall were in synchrony
with the waves. When a roller passed overhead you rose with it, negating its effect. But *Abex* was too big to rise and fall in time with the swell.
The bow was in the air when a trough passed under it and buried deep
when the next wave came, doubling the problem. I looked around the
barge, the dozens of specialists, the banks of generators, racks of gases,
and decompression chambers, all primed for work on this windless,
sunny day. None of it was of any use.

With this equipment, there was nothing Dilip could do. Heave-and-
pitch compensators would stabilize the stage, but we didn't have them.
"Like I've always said," he shrugged, "to do this job properly we can't cut
corners. It's got to be the right kit."

LIKE HEAVY FOOTFALLS felt before being heard, the long, low swell
that prevented us from diving was ominous. When Dilip reported our
problems over the phone, Ong remained determined. It seemed to me
that for him to give in would mean he had been defeated by Hatcher,
and the precious trust of his investors would be damaged. In the end
Ong decided that though money had been spent it wasn't wasted, be-

cause now he had a better idea of what was on the seabed. The problems merely required looking for a new approach.

Innovation and determination have always been essential ingredients in salvage. In 1922, an Italian commodore named Quaglia and his salvage team invented an entirely new diving system in order to reach the riches that lay in the carcass of the P & O liner *Egypt* lying on the bottom of the Bay of Biscay. Enclosed by Brittany in France to the north and by Spain to the south, the full force of the Atlantic becomes concentrated in the bay, while the shallow continental shelf extends far offshore, further amplifying the swell. Only a few months earlier, the *Egypt* had been rammed in thick fog and sank thirty miles into the bay. Quaglia sealed a man inside a narrow steel cylinder with three tiny portholes and dangled him from a crane into the water. Once he was four hundred feet below the surface, another crane lowered explosives, which the submariner directed into place on the hull. Quaglia and his team spent the next thirteen years in this way, gradually blowing their way through three decks and into the bullion room. In 1935 one of the divers finally guided the grab-bucket to haul the last of 180 gold bars and eight thousand sovereigns from the guts of the ship. Their perseverance paid off. The treasure was worth $2 million at the time of the *Egypt*'s sinking, the equivalent of more than $50 million today.

Spurred on by Dilip's insistent recommendations, Ong began to consider using equipment that had never before been used on an archeological project. *Rockwater 2* was a state-of-the-art offshore diving vessel. She didn't need anchors to hold her in position; she used "dynamic positioning," whereby thrusters at each corner of the ship were linked through a computer to a sensitive Global Positioning System. Feed in the coordinates of where you wanted to place yourself, and the automated system would do the rest. If the weather got too threatening, you just hauled up your divers and left. Once the weather cleared, you could be back on site with divers down almost immediately. There was no time wasted setting and recovering anchors. *Rocky 2*'s main advantage, however, was her saturation diving system. The divers would shuttle between the onboard

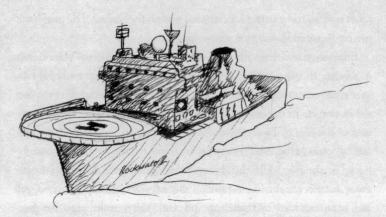

chambers and the seabed using a sealed diving bell launched from a "moon-pool." Invisible from the outside of the ship, the moon-pool provided sheltered access to the water from the vessel's center, where it was least affected by any pitching and rolling. To soak up any unwanted movement that remained, the winches for the guide wires were also fitted with heave-and-pitch compensators.

The photos of *Rocky 2* certainly looked impressive. Three hundred and twenty-three feet long, with a helideck spreading across her bows, the bright-red ship was a league apart from *Abex*. When Dilip first proposed his idea, Ong had laughed. There was no way he could afford a ship like that. But Dilip had discovered that the vessel was in the area and without contracts for the next month and a half. Once again working his industry connections hard, he negotiated *Rockwater*'s rate down from $87,000 to $38,000 per day. Ong put the idea to Mensun. The Hoi An project using saturation would be a very different kind of endeavor. Given the running costs, the excavation would have to be much shorter. While everything would still be methodical and controlled, the procedures would have to be streamlined; scholarly analysis would have to be done at a later stage. At the speed necessary to make *Rocky 2* financially viable, the work done on site would now be much closer to salvage than it was to archeology.

Nonetheless, using the expensive technology seemed the only way that the cargo would be recovered in anything resembling a disciplined manner. And it seemed far better than leaving it for the fishermen's rakes. Mensun knew that archeological purists might accuse him of legitimizing a treasure hunt, of lending his name as well as that of Oxford University to the desecration of a historical site, but he felt that academic scruples could not be allowed to risk losing forever the opportunity to recover and record such valuable material.

Proceeding with saturation meant Ong would have to raise the stakes dramatically. Of the three Malaysian-Chinese brothers whose wealth was behind all Ong's investments, only one had been involved in financing the Hoi An wreck project up until that point. To fund *Rocky 2*, the original budget was now vastly increased; the other two brothers were called in to help shoulder the risk. They trusted Ong. He had been with the family for a long time, and his gamble with the *Geldermalsen* had paid off handsomely. Nonetheless, this new plan was far more expensive and the returns were much less certain. The brothers knew a good deal about Chinese porcelains and their value. The Vietnamese versions seemed crude by comparison. They were of course rare, but they had barely been written about and were hardly collected. How could anyone be sure they would make a good investment?

In an apparent effort to convince the brothers of the ceramics' worth, Ong decided to bring Mensun to one of their meetings. Valuation wasn't his field, Mensun cautioned the brothers when he arrived. Judging the value of Vietnamese wares was beyond his expertise. He therefore emphasized their historical significance, allowing his audience to assume this would have a significantly favorable impact on the price they would fetch at auction. Despite his professed ignorance on the subject of Vietnamese ceramics, Mensun knew enough about the antiquities market to be surprised by the high figures Ong was projecting in his PowerPoint presentation. Ong maintained that they were all official prices, gleaned from sales at top auction houses around the world. Mensun later told me that it took him a moment to realize that Ong had found the most valuable of the occasional Vietnamese pieces that had been sold, estimated how many examples of that type might be on the seabed, and then multiplied the figures to reach the 100,000 pieces he hoped lay intact. The totals were impressive.

Mensun bit his tongue. He wanted to see the wreck studied, regardless of whether anyone made money from it. Though present primarily to confirm the archeological importance of the site, he was asked questions concerning the schedule, budgets, and feasibility of the excavation, as well as the use of saturation. Despite his doubts, Mensun backed Ong up. He had been in Ong's fundraising position before and he knew how delicate things could be. The response from the brothers was cautious head-nodding. In the end, they agreed.

For Ong it was double or quits. So far he had been cautious in the salvages he had organized himself, employing only those techniques and equipment whose efficacy he had witnessed first-hand. Now he was about to embark on something bigger than anything anyone had attempted before.

With the funds at last secured, it was Ong's turn to have second thoughts. He was rushing into the venture. It was late in the season and the weather had already shown its teeth. The long swells that had prevented the divers from decompressing on an open stage were a premonition of distant typhoons. If they chose to bear down on the Dragon Sea, even *Rocky 2* would be unable to work. Ong decided to close down the project and reconsider.

Mensun had no option but to go along with Ong's decision. The fragile combination of circumstances that had come together to form the excavation had fragmented. Though Ong talked as though he wanted to try again, Mensun held out little hope that there would be another season.

I felt philosophical about it all, perhaps because I could afford to be. The adventure, exploration, and discovery that I'd so looked forward to had turned into long weeks of waiting. We'd merely glimpsed what we had come for, nothing more. The Dragon Sea was beginning to seem like a canny temptress, flashing a leg to lure us close and then brushing off our clumsy advances with disdain. There was no farewell ceremony, no thanking the Sea Emperor for his clemency. We simply left.

Sitting on *Abex*'s port side, my legs swinging above the water, I watched the Cu Lao Cham islands slide past. The breaking dawn joined the swell at our backs as we retreated. The fishing fleet filed past, forced close by the island channels, the languid beat of their engines pushing their smooth hulls through the water. From every wooden cabin pairs of eyes stared up at our floating steel island, its deck bristling with high-tech equipment. Balanced barefoot on the narrow gunwale of his boat, a young fisherman locked eyes with mine. His cap and clothes had been weathered by sun and salt to match the faded tones of his sampan. The hard look in his eyes became tinted with amusement as we passed. Despite all our fancy equipment, we had failed. His head was still turned my way as he disappeared into the glaring reflection of the morning sun.

CHAPTER 10

A Second Chance

ONG HAD NOT given up, however. He had simply decided to wait until the typhoon season was over before relaunching his assault on the Hoi An wreck. What's more, he had realized that employing *Rocky 2* was not the only way to bring a saturation diving system to the site. Though Dilip still insisted that using a dynamic-positioning vessel would prove more efficient in the long run, he had conceded that the job could be done with a barge-mounted system. The weather window did not open again until April, giving him and Ong a few months in which to rent a saturation barge from one of Singapore's dive operators. This could be anchored to the seabed at four points, just as *Abex* had been, but, by using a proper tug equipped with an anchor-handling winch, they could avoid some of the problems of the previous year. Though it was not his ideal solution, Dilip had got his way; he was going to be able to work with a saturation system. What he did not bank on was Ong's finding a still cheaper option.

After the end of the Vietnam War in 1975, a Norwegian aid agency donated a saturation system to the Vietnamese as part of an international recovery program. The Norwegians had tried to train the new owners in operating the system so that they might use it in the oil fields off Vung Tau, but the ruling Revolutionary Committees saw all Western advice as Western influence. Once the Norwegians had left, the newly founded national oil company Petrovietnam made some trial pressurizations but suffered several near-fatal accidents. The system fell into disuse, rusting away in a warehouse while parts were slowly stripped away. There it remained for the next twenty years, until the day Ong told VISAL that he was looking for a cheap saturation system.

When VISAL mentioned the Norwegian system, Ong commissioned a Singaporean diving contractor to inspect it, hinting that if they decided it was possible to resurrect it they would stand a good chance of winning the contract to refit and operate it. When the company emissaries returned from Vietnam, they reported that though the equipment lacked many safety mechanisms, the main pressure vessels would hold and the ancillary equipment could be refurbished. The apparatus would not be legal in any of the world's oil fields, not even those in the anarchic South China Sea, but this was not an oil contract and neither Ong nor the contractor was bound by offshore regulations. It was "a pig of a system" according to those who later ran it, but it would be several times cheaper than the alternatives.

Dilip was incensed. Once again he had committed himself to the project and then had the rug being pulled out from under him. The oil company Elf was courting him, offering him more than $1,000 a day to run the recovery of a shipwreck located on one of their oil fields. He had given his word to Ong, but he was feeling tempted by Elf's offer. His main worry about the Hoi An project was safety. Once the barges left Singapore they were his responsibility, not Ong's. With men locked into saturation chambers on a platform without its own propulsion, he had to be certain that he could look after them if the weather closed in. That meant renting a proper tugboat. When he discovered that Ong was looking at another cut-rate Indonesian vessel, Dilip took matters into his own hands. While Ong was away he contracted the type of tug that he was used to working with. *Ena Supply 3* would cost $2,500 per day (compared to $750 for the Indonesian alternative), but she was an ocean-going workhorse with anchor-handling winches and a trained crew. To Dilip she was worth every penny. When Ong found out he was furious, Dilip later told me, but by then the deal had been made.

IN MARCH 1999, I returned to Singapore's Pandan Road dockyards. MARE was now engaged in what was set to be the most expensive underwater archeological excavation ever, and I'd been hired as the archeological manager, charged with fitting Mensun's objectives into Ong's operational strategy. This was the future with MARE of which I'd always dreamed. I was to start putting together what was needed for the archeological side of the operation while the saturation equipment was mounted on the barges.

Abex was in her familiar berth in the dockyard, but looked forlorn. She had been stripped back to her bare accommodation. Standing on her deck with a blowtorch was a sturdy-looking individual, bare-chested in a pair of denim shorts and boots. "2-REP" was tattooed across his thigh, while other tattoos were blotchy and indecipherable, evidently done with a sewing needle and ink. He watched as I walked up to introduce myself, and then told me his name: Nigel Kerr. That figures, I thought. MARE team members who'd worked with Kerr on the *Nassau* had filled me in on his background, and he was as intimidating as his reputation. Ong had lured him back from his security work in Algeria, wanting Kerr to be his man in saturation and keep control of things on the seabed. It sounds exaggerated now, but at the time it struck me that his opaque blue eyes and blunt nose were like those of a great white shark. Somehow when he spoke, his words were backed up with a physical confidence that was beyond certainty.

On a nearby causeway a much bigger barge had been winched out of the water and loomed overhead, "Tropical 388" spelled five feet high on her bow. We walked up the long gangplank and Kerr took me to meet the master of *388,* the barge's equivalent of a captain. Hartmut Winterberg shook my hand with an enormous fist and in a clipped Austrian accent welcomed me on board his barge. He was in his late forties and a mane of white hair reached the base of his thick neck. When I asked how work was progressing he pulled out a pair of half-moon glasses from where they lay suspended on his chest, perched them on his nose, and began to read from a list on a business card–sized piece of board that he had pulled from his back pocket. "Today we put in the galley, the stairs. Secondary power systems tomorrow, toilets . . ." The list went on, all scrawled in tiny writing. When I asked why it was so small, he told me he hated clipboards.

Although there was little over a month to go before the diving was to begin, the saturation system was still scattered over the bare deck of *Tropical 388.* Dirty, scuffed orange spheres lay inside rusting white frames, technicians bustling around them. Showers of sparks streamed from angle-grinders and blinding white fireflies floated around the arc-welders, all to the accompaniment of a deafening hollow clanking. All over the deck squatted Indonesian workers, heaving lump hammers overhead and pounding them into the flaking deck, methodically beating rust from the hull.

Looking around and peering inside the disconnected chambers, I recognized a long black ponytail as that of Jack Ng, the tattoo-covered diver on *Abex* the previous season. Sweat dripping off his nose ring, he showed me around the rounded steel capsules. They were roasting hot in the tropical midday sun even though all the hatches were open and the porthole glass had been stripped off and not yet replaced. Rusting bolts protruded from some of the fittings, while bundles of decaying wire sprouted from other ports. Jack showed me the four main chambers: a main habitation module containing four bunks; a secondary bunk room, with room for three at a squeeze; a transfer-under-pressure chamber through which the divers would pass on their way to the last chamber—the bell itself—which would ferry the divers to and from the seabed. None of them was linked up yet and all looked a long way from being ready for anything, let alone keeping six men alive at eight atmospheres of pressure. The scene resembled a small-town chop shop knocking together a space station. I was fascinated by the nonchalance of the crews as they threaded bolts onto hatches. One mistake—a worn seal or valve, damage to an exposed high-pressure fitting—risked a "Massive, Catastrophic Decompression," or MCD, when the full force of the system's pressure could explode through an opening. The gas inside the chambers—including any contained in a man's lungs and sinuses—would expand many times over in a fraction of a second. Off the coast of Spain in the mid-1980s, an inexperienced crewman on deck began unbolting the wrong clamp on the outside of a saturation chamber, loosening a hatch that was under pressure. The door had blown off with such force that it killed him and two other deck hands, as well as all those inside. There's not much gas inside the human body—the fact that it is 90 percent water enables it to withstand huge pressures—but what little there is does catastrophic damage during an MCD. It explodes from the chest, rupturing the lungs if it is unable to escape the throat. Gas bursts from the sinuses, ripping open the eardrums. Vapors trapped inside the bowels expand, though the intestine is usually elastic enough to absorb the internal explosion.

Amidst all this trauma, it's difficult for doctors to say what exactly kills a human during such violent decompression. Post-mortems are complicated by the gas that continues to fizz out of the body tissues. Bubbles are found within the heart and main arteries but all could have been formed after death, as could the widespread subcutaneous emphysema—a ballooning of the skin beneath the neck caused by trapped air, a condition otherwise mostly found in knife or gunshot victims. Instead medics point the finger of blame at a much more insidious cause: free-floating fats that are found surrounding bubbles in the heart, pulmonary arteries, and the brain. Though exactly how and why they are unleashed remains mysterious, what's certain is that explosive decompression is not a pleasant way to die.

Robbie Robinson was busy in the "control van," a shipping container converted to hold the monitoring and control instruments for running the system. As one of the two life-support technicians, his job was to make sure that the divers survived their ordeal. The gauges in the van were all dials and the valves were all manual—without a single digital display, the panel looked like a set from a World War II submarine drama. Tracing wires and piping from the back of the panel, Robbie whistled away. He'd been in the business since the 1970s and was enjoying the challenge of putting the system back together. At least it was a ground-up refit, not just a patch-up job, he reasoned. What's more, he had managed to get his son on board with him as deck support.

Chad had been running a successful restaurant back home, when one day he announced that he was quitting. He wanted to be a diver like his dad, he said. Robinson senior was half flattered, half despairing, but he also knew better than to try to discourage Chad. Robbie had helped engineer the stroke of luck that saw them working together, and it was good to have his son on board. This way he could keep an eye on him, although there seemed little need. Chad appeared to be taking to the work and his workmates well. He had the right attitude. He was tough and pragmatic, and even refrained from bragging about his job when he was back home. When people asked what he did for a living, Chad told them that he sold ice cream to avoid the inevitable reaction he would get if he admitted to being a deep-sea diver. Though his son had done one saturation dive and wanted to do more, Robbie was glad—with a system in this condition—that Chad had come on this job only as deck support. That said, Robbie was there to make sure everyone inside the system survived, family or not, and that was what he intended to do.

Aside from avoiding catastrophic decompressions, one of the most important things to watch was the temperature inside the chambers. Helium acts as the opposite of insulation, instead coupling a body tightly to the temperature of its environment. The chambers needed to be kept close to body temperature or those in the helium atmosphere inside would either freeze or cook. The Environmental Control Unit was what they would all rely on. These units were notoriously temperamental, and Robbie didn't like the look of this one. The problem of temperature regulation intensified with pressure. In normal conditions an overheating person will begin to sweat, then become nauseous and disoriented as heat exhaustion sets in, before succumbing to potentially fatal heat stroke. Under pressure the situation is different. At anything above thirty atmospheres, the body loses its ability to regulate body temperature. The body does not sweat, and it skips the formality of heat exhaustion. Heatstroke strikes without warning.

Diver Roland Dashu knew all about heatstroke. Though he wasn't fixing the Environmental Control Unit himself, he had worked in the casualty ward of a major hospital in Shanghai for the past three years as a healthcare assistant. He'd left the diving industry to placate a girlfriend but as time went by, his memories of working in saturation had softened. Eventually the thought of all the money he could make had lured him back. This would be his first time inside the chamber since returning to the industry, but he wasn't worried. He was looking forward to it in fact, though because he did not want to lose the confidence of his fellow divers, he had bluffed his way through their questions as to where he had worked most recently.

Even before his time at the hospital, Dashu had known that saturation diving was preferable to decompression diving. Decompression places great stress on the body, and for a professional who dives two or three times a day every day, the consequences can be severe. Aside from the immediate risks of embolisms and the bends, over time microscopic bubbles can collect in the bones, eventually causing them to rot away from within. Saturation is kinder, but only if everything goes right. Unfortunately, the time it takes divers to decompress is nonnegotiable. The divers must remain pressurized even when the vessel that is carrying the system gets into trouble. All modern systems are now required to have a Hyperbaric Rescue Vehicle (HRV). A chamber encased within a lifeboat, an HRV allows divers to escape a sinking ship while continuing a controlled decompression. Several incidents prompted the regulation, but none were as chilling as the case of the saturation barge *DB-29*, caught by a typhoon in the South China Sea off Taiwan in August 1991 with four divers "in the pot." The *DB-29* was running for shelter when two anchor buoys lashed to the deck worked loose and ripped open a hatch. Water started pouring into the hull faster than it could be pumped out and the barge began to keel over. Emergency procedures to decompress the divers were started, but their bodies were still days away from atmospheric pressure. Two agonizing hours later, the barge flipped over and eventually sank to the seabed two hundred feet below. Trapped inside the chambers, the divers had no chance. Even if they survived the cold, the gas remaining inside the chamber would keep them alive for only two days—a long time, but nowhere near long enough for a suitable rescue vessel to reach them.

Three of the four divers decided to confront the laws of physics and physiology rather than face the psychological torture of waiting for death. After writing letters to their wives and loved ones, they flooded the transfer-under-pressure chamber and opened the external hatch, making a break for the surface. Their contorted and ruptured bodies were found floating among the debris of the stricken barge, letters taped to their legs. Eighteen other bodies were recovered from the water that morning; the body of the fourth diver was not among them.

A year later the dive company returned to the site of the sinking, hoping to salvage the dive bell and other valuable pieces of equipment. When divers reached the seabed and peered inside the saturation system, they made a grim discovery. Reclining inside was the body of the fourth diver. Bryan Shepherd had rigged a hammock and stretched a hose to the uppermost corner of the chamber, where the last oxygen would remain. Resigned to his fate, he had decided to die as comfortably as he could.

On *388*, the dive contractor had decided that the secondary bunk chamber could double up as a cheaper version, dubbed the Hyperbaric Rescue Chamber (HRC). Without its own propulsion, the idea was that it could be sealed and cast off the ship in the event of trouble. Dashu had no reason to question the fact that the long section of trunking connecting the HRC to the main chamber was attached with normal bolts, rather than the explosive bolts that would normally be used to ensure a quick release. He didn't stop to examine the foam that surrounded it, a new coat of orange paint filling the holes that had been torn from it in the past three decades. Nor did it cross his mind to check if any flotation tests had been done.

CHAPTER II

Blowdown

A MONTH LATER, off the coast of central Vietnam, Mensun, Ong, Dilip, and I stood at the end of a long, narrow cabin watching the dive crew file in, scowling beneath their grease-smeared faces. They wore stained overalls and big, scuffed boots. Work was continuing on the saturation system, though pressurization, or "blowdown" as it is called, was scheduled to start in a few hours. The gray walls and fluorescent lighting in the converted shipping container brought each man into sharp focus as he stepped in. They were all squinting and seemed hunched over, as if they were being brought in for interrogation. The door shut.

"Right. That all of you?" said Dilip. There was a show of looking around, checking, and a mumbling of assent.

Mensun coughed. "Gentlemen, welcome. We're all very anxious to get started so I'm going to keep it as brief as possible. I know a lot of you have worked on heavy industrial jobs, that some of you have done steel-wreck salvage, and some may have even worked old ships before—"

The door opened, and Mensun paused. A middle-aged man stood there, one hand jammed into a pocket of his shorts. A baseball hat was perched at a comical angle on his head, shading bloodshot blue eyes.

"This the women's institute?" he asked in a deep Australian drawl. Several of the divers guffawed before catching themselves.

"Yes, that's right, Tony," said Dilip. "Want to sit down and we can start?" I remembered Tony Turner from the dockyard. He was one of the two supervisors who would act as middlemen between the archeologists and the divers. I watched him shuffle into the corner, pushing aside one of the others with an elbow. Out of the corner of my eye I searched Dilip's face for signs of irritation, but found none.

"Right," Mensun continued. "As I was saying, I know that we have a range of backgrounds and expertise gathered together for this project, all of which I'm sure will come in very useful. However, I also know that no one's done a job like this before, for the simple reason that there's never been one. This is the first time that anyone's used a saturation system to do an archeological excavation—which makes this the deepest excavation anyone's ever attempted, not to mention the most expensive."

A barely suppressed snort came from the end of the room. Mensun continued without a pause.

"It's going to be a very technical operation, one that's vital we get right. You'll be handling extremely fragile pottery here—pottery that's also extremely valuable. I'd just like to fill you in on a little of the back story here, to give you an idea of what we're dealing with."

Mensun began to outline what little we'd been able to discover about the shipwreck during the previous season's work. We knew, for instance, that the vessel's main cargo appeared to have been pottery from northern Vietnam. He mentioned that the ceramics were rare, so much so that no one knew for sure exactly when they had been made. It was a difficult pitch, trying to convert the room full of hardened oilmen to the cause of scholarship, and I could see that Mensun was treading carefully along a well-considered path.

"If this wreck gives us what we hope, it will answer a lot of questions," he said. "Each and every piece is of huge importance for the archeology, and could give us vital clues. So please, take it easy down there. And I'm sure Ong will thank you to as well."

Ong smiled. "Yes. I will thank you. Remember—I'll be able to tell if you do break anything. I'll feel the ache in my balls." A few of the men smiled and laughed politely.

Mensun took back over. "Now, although the pottery is our number-one priority, there are other things we want to look at as well. The ship itself, and the personal possessions of her crew and passengers," he said, before describing what he expected them to find. We knew the vessel was a junk of some sort, simply because no other ships of any size were sailing in the area in the period that the pottery was manufactured. He told them that very few other junks had been properly excavated, and emphasized that we were all being handed an enormous responsibility, for there were important questions that remained unanswered about how these enormously successful, flat-bottomed vessels had been built. Mensun was hitting his stride as he spoke about the structure of the ship. I thought of all the times I had seen him behind the podium in car-peted lecture halls. They felt a long way away from our present situation.

Initially Mensun had wanted to brief only the divers, but Dilip had advised us that it was better everybody was told the same thing at the same time. The deck crews needed to know what the divers were likely to want or need, and vice versa. Keep it simple, he had added. We would be better off laying out the basics and letting the divers learn on the job, guiding them through the supervisors.

"You're all seamen and can appreciate that the watertight bulkheads characteristic of junks should make the vessels all but unsinkable—it's the same idea that Victorian engineers claimed to have invented when designing the SS *Great Britain* and later fitted in the *Titanic*. Which leads me to another question. If the bulkheads were watertight—as Marco Polo once claimed—then what did sink this ship?" Mensun paused, as though considering the possibilities. He had snagged his audience now. Legs stopped swinging. All eyes were on him.

The dive crew consisted of a mixed bunch from Australia and New Zealand, Singapore, China, America, and the UK, but they would have to operate as closely as a family. The patriarch of the operation, known only as "Spanner," was from the city of Birmingham in England. He was the dive contractor's most senior figure on site. His large gold-rimmed glasses had steamed up with condensation from entering the air-conditioned office. His head was covered with orange stubble, matching patches of the Hawaiian shirt that hung down from his belly like a curtain, beneath which sprouted two thin, freckled legs. The two dive supervisors were Garth Sykes and the latecomer Tony Turner, both of whom would also play a paternal role, standing up for their divers but also expecting their orders to be followed to the letter. The life-support technicians—one of whom was Robbie Robinson, whom I'd met in the control van— would mother the incarcerated men, tending to their needs, wishes, and whims. The divers themselves, stripped down to shorts and T-shirts in readiness for the blowdown and wearing the defiant look of condemned men, were the breadwinners of the family. Everything had to be geared toward helping them. The deck techs like Chad Robinson would be assisting on the surface, lowering tools, repairing equipment, and doing any other odd job that presented itself. They were all experienced commercial divers as well, but a rung down from the saturation divers. They hoped to earn the wages of the saturation divers themselves eventually. In the diving world everyone had risen through the ranks, and most had trained in some manual trade before being admitted to the clan.

"Aside from the ship and her cargo," Mensun resumed, "we are interested in the men who sailed her." Any personal possessions that we found would help tell the story of both the ship and the world they came from. He cautioned the divers to be on the lookout for unusual objects, not just the pottery. Just about anything could turn up on the site—that was part of the drama—but among them we expected ink pots, weights for measuring goods, combs, buttons, medical equipment, even shoes. And coins. Coins were especially important as they would help us pin an exact date on the ship.

"We're going to need you to be our fingers and eyes," Mensun continued. Watching the screens would be no substitute for being on the seabed. However, from the surface we could keep an overview of the
situation as it developed. Communication between us and the divers
would be vital. Mensun explained that we would pass daily reports into
the chambers and that, in return, we needed the divers to respond with
sketches and descriptions of what they'd seen once their long shift in the
water was over.

"As you all know, this is a commercial excavation," said Mensun.
"Some of the ceramics will be sold to pay for all this equipment and professionals such as yourselves. But what we need to remember is that by
getting the full story of the shipwreck, by learning about the ship and
her crew, we will be adding not only to the value of whatever is recovered, but also to history."

This was classic Mensun. He had managed to unite the group behind
a quest that must have seemed as alien to them as the world of sub-sea
oil extraction did to me. The atmosphere in the room was as charged as
if we were a football team about to run from the locker room into a
packed stadium. Even Tony Turner had stopped snorting.

It was my turn. I stepped forward. I was to run through the practicalities of how we intended to accomplish archeology with divers who until
now had used a trowel only for plastering walls.

"We've got two priorities on the seabed," I began. "We've got to recover as much of the cargo as possible, while keeping track of exactly
where it came from. In order to help us build up this map of the wreck,
we need to be organized about how we recover. As you'll have seen, out
on deck we've got a lot of steel grids. Our first step will be to lay these
grids over the work area."

I ran through the plan. Each of the four-meter-square grids was divided into four sections, each a square with sides two meters long. They
would be laid level using adjustable legs, each flush with the sides of its
neighbors. We'd start at a single point on the wreck and expand outward,
laying more grids as we progressed. In the set sequence of procedures to

excavate each grid the diver would initially move his helmet camera over the contents, allowing Mensun or me to sketch a small map prior to its being disturbed. The covering of mud and disturbed material would then be removed with the help of an airlift.

"Once we've got a clean view, we're ready to scan," I announced. From beneath one of the tables I rolled out my *pièce de résistance:* an underwater camera dolly. The wheeled chassis, bright yellow, cradled a large underwater video camera that pointed straight down, while flexible lights arced over the top. I'd designed it to scan the wreck one grid at a time by rolling slowly along a set of tracks that fitted on top. The camera would record continuous high-quality video, from which I would take images to make up a photomosaic of the grid, and eventually the whole wreck site, layer by layer. If it worked, I hoped that we'd be able to dispense with Mensun and me having to sketch every grid by hand. It would give us better information faster, I thought, and therefore more cheaply. For that reason Ong loved the idea, and it fitted with my self-appointed role as modernizer of MARE. Only Mensun was skeptical, but I was sure my equipment would prove him wrong.

"Err, d'you aaarchiologists understand how it's going to be down there?" It was Turner's Australian drawl. "It's going to be a facking nightmare of ropes and cables. You've got your bell wires, diver's umbil-

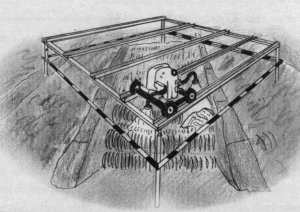

icals, bell umbilical, airlifts, recovery lines, anchor lines, markers. Currents'll be running and you're going to lose your pretty little camera soon as the job starts." He said all this staring at the floor, looking up only when he had finished, his opaque eyes staring blankly at me.

"Well, we can't lose the camera," I said. "Perhaps you think we should be recovering it after each scan to keep it out of the way?"

"You could do that, but you'll probably smash it 'gainst the side of the baaage pretty good every time you recover it." He raised his voice at the end, looking around the room at the other divers, who were grinning but avoiding his eyes. "And your pressure seals won't like it much. You might as well leave it on the bottom if you're going to do that."

"But we'll have to change tapes and batteries . . ."

Turner wasn't listening. He scratched at the bushy hair on his shoulders and said, "I tell you what. Why don't you just make things simple and tie it to the grid when you're not using it?"

"Gentlemen . . . , gentlemen . . . ," Mensun interjected with a smile. "This is a detail you can talk about in the next few hours. Let's keep things on track and talk about main procedures." He was right. Swallowing my irritation, I continued, addressing the others in the room.

After the video scan the diver would begin to unpack the first layer of pottery from the grid, loading it into a numbered plastic basket the size of a milk crate. Up on the surface, Mensun or I would record which plastic basket carried ceramics from which grid. The plastic baskets would then be put into a large metal recovery crate, which would eventually be winched to the surface by crane. If anything particularly significant or delicate was found, a separate padded container—the "specials basket," as it was called—would be used to bring it to the surface independently.

"Right," Dilip piped up when I'd finished. "As you all know we don't have long to finish this job. You've all agreed to twelve-hour shifts in the water. I'm sure you can all handle it." He looked around the room, as though challenging the divers one by one. This was four hours longer than a standard dive-shift while in saturation. Because of their high running costs, saturation systems always operated with divers working twenty-four hours a day. Usually this meant using three teams of divers, giving each team an eight-hour shift on the seabed. To cut costs, Ong had decided to reduce this to two teams working twelve-hour shifts. Dilip warned that the divers would tire quickly but did not pursue the matter further. That was an issue for the dive contractors, not him.

"One word of warning," Dilip continued. "This is an archeological recovery, not a salvage job. While we'd love to be able to give everyone souvenirs, we can't. It's against the rules. Don't try to help yourselves, either.

"Right. We break for lunch, then blowdown is scheduled for fourteen hundred hours—that's in two hours. Enjoy the fresh air while you can, boys."

I STEPPED OUTSIDE after the others and walked to the edge of the deck. The sun was hot, the sea smooth and flat. Clustered on the open water like an ocean-bound Las Vegas, the expedition vessels seemed incongruous. Two large tugs stood off from the barges, their superstructures imposing yet off balance, like truck cabs without trailers. An old patrol boat belonging to the project's police escort traced an arc around the flotilla, low-slung and predatory. A relic of the Vietnam War, a heavy machine gun was mounted on the bow. I half-expected to see Martin Sheen leaning against the cabin as in *Apocalypse Now,* squinting through cigarette smoke as he pondered his mission to kill Kurtz.

Abex was dwarfed by the new barge, *Tropical 388,* to which it was tethered. Only the living quarters and the archeological office were on board; the rest of the *Abex*'s deck had been cleared to make room for storage. *Tropical 388,* meanwhile, resembled a long, low oil rig. Winterberg, the barge master, had been busy. A dense block of corrugated shipping containers, mesh walkways, and railings made up the living space at the stern. The rest of the barge was dominated by a bank of gas cylinders, and just behind them was the massive latticed arm of a mobile crane. Welded to the deck, the crane's metal tracks were just discernible through the winches and piping that ran down the decks. Its skeletal

arm swept backward, the hook tethered amidships. Nestled beneath, encased by a sturdy white frame, were the outlandish orange spheres of the saturation system.

Most of those on board *Tropical 388* had now gathered outside the spheres. I watched the divers mill around. They had already stowed their personal effects inside and were posing, sour-faced, for some publicity photos with Mensun and Ong. Without exception they were physically large and powerful men, each an alpha male.

Nigel Kerr had been designated "chief diver." Normally all divers in saturation have equal status, but Ong wanted to have someone he considered his own man to be in control on the seabed. Having salvaged six

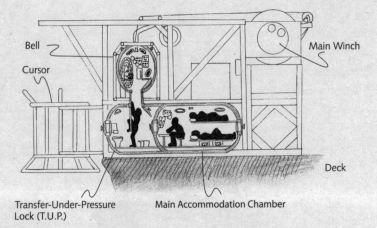

Bell

Cursor

Main Winch

Deck

Transfer-Under-Pressure
Lock (T.U.P.)

Main Accommodation Chamber

wooden wrecks in the previous few years, with various degrees of archeological involvement, Kerr knew and understood the work. But he had
never worked in saturation before—though he had completed his certification six years before—and his lack of saturation experience, coupled
with his being named to the unusual post of chief diver, had raised eyebrows among the dive crew.

Hector George, for one, measured respect for others mainly by how
much time they had clocked in saturation. I'd already heard about
George's notoriously short fuse. A New Zealander with half-Maori
blood and a permanent scowl, he had spent eight years in the SAS. If
that hadn't given him enough time staring into the face of death, the oil
industry had. He had been on a dive vessel in the Persian Gulf when it
had been hit by an Exocet missile, but his closest shave was on the ill-
fated *DB-29*. George was supposed to have been one of the four divers in
saturation, but a flight delay had forced him to miss the blowdown.
Someone else had taken his place. Working as surface support on deck,
he had been powerless to help as the barge turned over with the divers
still inside the chambers.

George glowered from beneath his heavy forehead, his massive arms
folded across his chest as he leaned against the chambers inside of which

he would soon be locked. The antiquated system had been built long before the DB-29, and had no hyperbaric lifeboat incorporated into the design. George had little faith in the small spherical chamber that had been designated as an escape chamber. No one had ever tried using one, and this looked like a token affair. There was no launch mechanism, nowhere for a life-support technician to sit on the outside nor any way for the divers to control the small amount of safety gas that was lashed to the outside, and not much flotation. If the barge went down, the divers would most likely go with it.

The lithe Chinese diver Roland Dashu was the last of Team 1. Though he was the smallest of the six divers, he made up for it with his swagger. His scalp was freshly shaved and flashed in the sun, his head twitching bird-like as he monitored his surroundings. The three divers who would make up the second team were Jack Ng, the Singaporean-Chinese diver who'd worked on the site the previous year but who was on his first saturation, and two old hands from New Zealand, Bernie King and Mike Hughes.

Mensun and Ong stood watching the men prepare to go through the hatch. Their ambitions depended entirely on these untamed, unkempt individualists. It was hard to tell who was more nervous—Ong, hoping that they'd be quick yet gentle enough to recover a large quantity of the delicate ceramics intact, or Mensun, whose frown deepened at the thought of these rough specimens being his eyes and hands on the seabed. The only ones who didn't seem nervous were the divers themselves.

Kerr, inscrutable, stepped in through the waist-high hatch with a casual nod. Then George, still scowling, squeezed himself through. Three others followed. Last was Dashu, who swung himself in with a whoop. "I love this shit!" he shouted, and pushed the hatch door closed.

INSIDE THE CONTROL VAN, from which all diving and chamber operations were directed, sat Robbie Robinson, one of the two life-support technicians. He kept the divers talking as the hatches were sealed, simultaneously checking the banks of gauges that were arrayed in front of

him. Robinson had worked every job on a dive vessel, but this was the
one that suited him best; he understood what the divers needed when
they were inside and knew the importance of keeping them happy.

Once he was satisfied the hatches were locked shut, Robinson
cracked open a series of valves, his hand poised while he watched the
gauges, ready to shut a valve off if anything went awry. Slowly the pressure built. Gas hissed quietly through the tubing that fed from the huge
banks of cylinders, through his saturation control panel and into the
chambers. It could even be heard entering the chambers via the micro-
phones that broadcast sounds from inside onto speakers in the control
room. Video screens were arrayed along the top of the saturation control panel showing feeds from cameras that monitored every part of the
system.

On the grainy black-and-white screens, Robbie watched the divers lying on their bunks, arranging their few possessions—books, photos, magazines. They moved little, except for taking occasional swigs from a water bottle. Like air being pumped into a tire, the air was heating up as the pressure mounted. It wasn't always so cushy. In the early days of saturation, when Robbie had been diving, salaries had been as outrageous as the risks. As soon as you were locked in, your pay became astronomical. For that reason, the supervisors would blow divers down in the bell: With less volume to fill the pressurization was quicker. It was the divers who suffered as the rapidly mounting pressure not only heated the gas to roasting temperatures but also pushed the fluid from their joints, making them crack until the fluid could force its way back in.

Once the chambers had reached two atmospheres, the equivalent of being under thirty-three feet of water, Robbie flicked a switch and spoke into the mike that jutted from his headphones. "Switching you to heliox now. Talk to me, let me know how she comes."

The divers sat or lay on their bunks and kept up their monosyllabic conversation. A low whistling sound began in the distance and got closer and closer, rising in pitch as the helium surged through the piping from the gas quads outside, its lower density creating a pressure front. Eventually there came a soft crump as the gas entered the chamber. The divers' banter started to change tone, the grunts turning to squeaks, as helium filled their voice boxes. Their voices would remain that way for the better part of the next sixty days.

CHAPTER 12

The First Haul

THE CLAMP SEALING the dive bell's bottom trunking was released with a dull, metallic clank and a sigh as the last remaining gas escaped the air lock. The winch motor groaned as it heaved the bell off the system, the orange sphere rocking as it dangled from its guide wires. Once it was clear, a different motor engaged and the bell began to track over the side of the barge until it swung above the white conical frame of the guide chute. Support technicians stood braced inside the frame, ready to fend off the bell to protect the racks of emergency gas or other external fittings from damage. Then the winch groaned once more, and the bell began a slow descent. As it entered the water I searched for a glimpse of the divers crammed inside, but its one upward-facing porthole was too worn to see through. Soon the symmetrical white lines of the frame were distorted into broken curves. I followed the shrinking orange and white shape with my eyes until all I could see were the guide wires glinting, running straight downward like railroad tracks heading into the night.

Like the diving helmet, the basic principle of the diving bell hadn't changed that much since 1691, when Edmund Halley (of comet fame) invented his wooden prototype. The trapped air in the bell provided a dry environment for a diver to sit in while allowing easy access to the water through the opening at the base. Halley's bell was kept upright with a skirt of lead and held in position with weights in the same way that the modern version relies on a "clump weight," though today this also serves as an anchor for the wires that guide the bell's ascent and descent. Halley's method of getting air to the bell was to lower it in barrels, while now it is pumped from the surface through a hose. To supply the diver, Halley used a simple leather hose that led from the bell, held by the diver's assistant. Again, a similar technique is still used in modern bells, with the bellman controlling the gas supplies of his colleagues in the water.

There, however, the similarities end. Halley's bell could operate only down to sixty feet, but in today's saturation bells it doesn't make much difference if your work site is at ninety or nine hundred feet. The modern version has lights, communication, and, more important, onboard safety gas. If the main umbilical to the surface is somehow severed, the divers have a few hours in which to pray for help.

I went back inside the annex to the main dive-control room and sat with Mensun, watching Garth Sykes prepare for the first dive. One of the two supervisors, Sykes sat next to Robbie Robinson, facing the wall of gauges, valves, and warning lights that enabled them to keep an eye on all aspects of the system. At his end of the array of monitors showing the bell and chamber interior were video feeds from the divers' cameras.

Sykes talked the divers through their equipment checks. Their voices were difficult to understand and I caught only the occasional word. A de-scrambler unit, designed to translate their helium voices back into the tones of normal speech, was fitted to the panel. Sykes was dismissive of it. On a deeper dive, when the vocal distortion is even worse, the old analogue device might have been useful; otherwise, it was easier to let your own ears do the de-scrambling. Interpretation got easier after decades of practice.

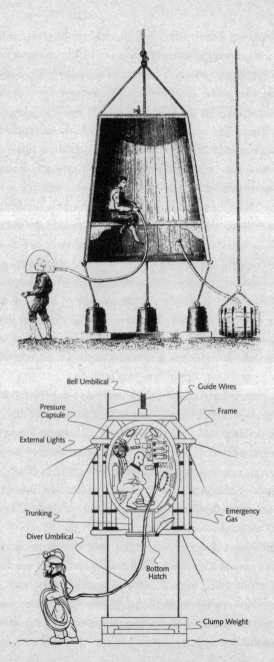

And Sykes had practiced for decades. He had been a boy when he bought a scuba set made out of old fire extinguishers with kitchen taps for valves and taught himself to breathe underwater, sitting on the bottom of the river mouth near his Australian home. Years later he began working in the oil fields of the Gulf of Mexico, where saturation diving was in its early stages, aiming to learn from the American divers there and hoping eventually to find underwater work. His experience on the riverbed stood him in good stead: When the rig's boss had asked if he was a diver, he could truthfully say that he was. Later that same day an old bronze diving helmet was lifted onto his head, and thereafter he worked as a commercial diver. Not for a long time did the offshore industry require certification (a response to the large numbers of divers who had died in accidents). When they did, a grandfather clause granted old-timers like Sykes their certificates simply on the grounds of their experience.

Sykes seemed the most universally respected of the entire dive crew; no one had a bad word to say about this sixty-two-year-old "gentleman of the diving industry," as he was known. Though he was a regular supervisor on top-of-the-line vessels like *Rocky 2,* the age and history of the system he was controlling now didn't seem to bother him. Like a mechanic who owns an old car, his intimate understanding of how the system worked meant that he rather enjoyed its glitches; they were part of its character. Besides, there was only so much that could go wrong; he felt a certain relief to be working with valves you could twist and pressure gauges you could tap. When something went awry with the modern systems, you needed a computer specialist to fix them, not a dive technician.

Two monitors were mounted in front of Mensun and me, one for each diver's camera. Looking into the dive-control room at other monitors in front of Sykes, the bell seemed crammed with bodies. Team 1 was inside: Kerr, George, and Dashu. On the bell camera, I watched Kerr putting on his helmet. Every movement required negotiation. He was already half in the water, standing on the bottom rung of the ladder inside the exit tube to make more space. George double-checked the fittings before reaching forward and removing the lens cap from Kerr's helmet camera. As he did so, the left-hand monitor on the desk in front of me burst into a full-color shot of George's surly jaw.

KERR MUST HAVE been relieved to be the first one in the water. Not only was the bell a tight squeeze for three people, but George was glowering at him as though waiting for him to make a wrong move. Eight years having passed since his course, the number of valves and gauges in the bell must have made it difficult to remember what was what. He checked his cylinder of safety gas one more time and lowered himself down.

Looking upward, the orange sphere and white frame of the bell filled his view, its lines sharp against the black water around it. Among the glaring exterior lights was a perfect circle of rippling green water through which he had just descended. The air in the bell kept the water lapping at the bottom of the open hatch as calmly as bath water, though shadows were moving across its surface. George was beginning to suit up. There was no sign of the wreck on the seabed a few feet below him, but that was no surprise. They had aimed to land off to one side of the site to avoid doing damage with the clump weight that kept the bell steady on the seabed. Kerr knew which way to walk when the time came. However, there were duties to perform first.

"Right, Diver One, you want to take us through the bell checks?" came Sykes's voice in his ear. There was a crackle of static alongside it; somewhere in the myriad of cables from the surface to the bell to his helmet there was a bad connection.

"Roger. Starting external bell checks," Kerr replied, with a chuckle at the sound of his own squeaking voice. He began to pull himself over the outside of the bell and check that nothing had been damaged or come loose during the descent. First he checked that the bell's backup gas cylinders were intact and secure, and then he inspected the first of the heavy U-bolts that held the bell to the guide wires. He hauled himself over the top of the bell to reach the other guide-wire mount, passing over the foggy glow of a porthole. Everything was as it should be. The checks finished, Kerr made sure the spiraling, multicolored strands of his own umbilical were clear. Then he jumped. He fell through the water, extending his arms and pushing his hands palm down at the last moment as though flaring a parachute for landing. His boots settled into the mud. He was back.

The water was darker than he had remembered from two years previously. Then again, everything about this visit was different. Behind him the bell hung above the seabed like some alien spacecraft, emitting beams of light that sliced through the blackness. He wasn't floating or kicking his fins to move but walking, and his breathing was easy. He wasn't going to have to leave in a minute's time; he had all the time in the world. But best of all he was clear-headed, not drunk out of his mind on high-pressure nitrogen.

THE FIRST BELL RUN was a vindication of everything Dilip had been pushing Ong to agree to for more than a year. Having the same divers in the water for almost twelve hours avoided the stopping and starting that had plagued our previous attempt. The barge was easily maneuvered a little closer to the wreck with the bell still on the seabed (with Kerr guiding its progress from the raised clump weight). The previous season's work site was quickly relocated; it was right in the center of the mound. The grid itself had disappeared, most likely caught in a fisherman's rake. Another grid was dispatched from the surface and set up on the same spot, quickly followed by three others that were placed around it. One of the three large recovery crates was lowered, each of its six compartments full of plastic baskets into which the pottery would be loaded.

Finally, airlifts were sent to the seabed. Long sections of pipe with an air inlet at one end, the airlifts functioned as powerful underwater vacuum cleaners. They were essential tools. Without them mud and sand

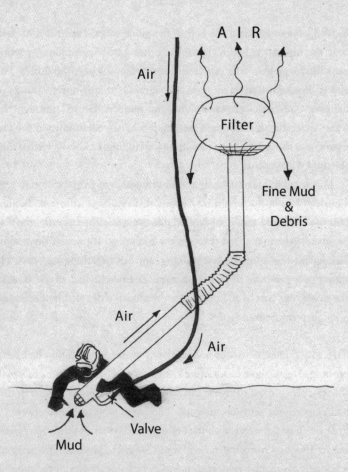

rapidly collapses back into a work area and clouds of sediment obscure the water. Air enters at the base of the pipe and rises upward, drawing water, mud, and sand with it. But small objects can easily be sucked up too and for that reason airlifts on a wreck site have to be used with extreme caution. Though a grating covers the inlet and filter bags are placed over the outlet these cannot be relied on to catch all very small finds.

Each armed with his own airlift, Kerr and George started clearing the mud from two different grids. Soon Mensun and I were frantically sketching the images from each of the monitors, straining to keep up but delighted at the progress that was finally being made. Two days later, the first crate was full and three of the ship's bulkheads had begun to be revealed. With saturation, we had already accomplished more in forty-eight hours than we'd been able to in two entire seasons without it.

An air of workmanlike productivity immediately settled over *388*. As soon as the divers had been locked into the chambers, life on the barge switched to a twenty-four-hour rhythm. Crews were split into two shifts, one running from midnight to midday, the other from midday to midnight. Of the forty-five personnel on board over half were directly involved in operating the saturation system, while the Indonesian deck crew bustled around them. Everyone (except of course the divers, incarcerated in the chamber mounted on deck or working on the seabed) came together at midday and midnight in the mess, where breakfast and dinner were served at the same time, while six o'clock in the morning

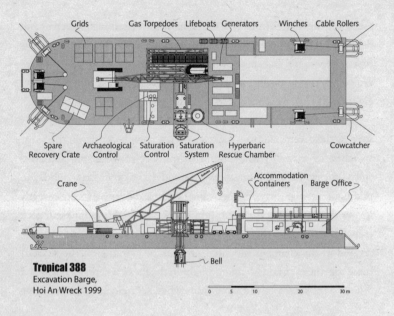

Tropical 388
Excavation Barge,
Hoi An Wreck 1999

was lunchtime for one shift, and six o'clock in the evening was lunchtime
for the other.

THE WHOLE DECK crew turned out to watch as the first crate broke the
surface, mud-brown waterfalls gushing from its grated sides. Mensun,
Ong, Dilip, and I stood among the stacks of grids on the foredeck,
squinting into the sun to catch glimpses of the white ceramics that
showed through the layers of blue plastic baskets and steel mesh. The
dripping crate swung slowly around the front of 388 toward the waiting
back deck of *Abex,* which had been drawn in close. Tag lines trailed from
each corner, dipping into the swell below. The crew waiting on *Abex*
reached out and grabbed them to control the crate's spinning and to
guide it in for landing. The crane's cab seemed level and stationary, but
the hook was rising and falling ten feet. Though a low swell was running,
the sea looked calm. Looking upward, I quickly realized the problem. The
crane's arm stretched up high, its tip moving fast against the clouds. At
that height, the gentle rolling of the barge was greatly accentuated, and

as a result the crate was swinging like a pendulum as well as rising and falling. The effect was accentuated by the decks of *Abex* and *388* moving independently on the sea. A padded landing area had been created out of tires and thick rope, but it looked pitifully inadequate.

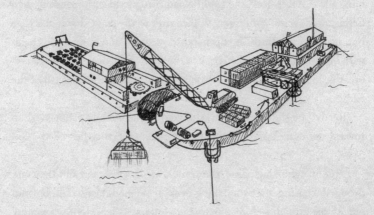

Suddenly I was terrified that the flecks of white were all that I would ever see of the crate's contents, that these ceramics had survived four hundred years of immersion and four years of looter's rakes only to be smashed on the decks of their supposed rescuers.

Hartmut Winterberg stood with his legs spread wide as he barked orders to both the crane driver and the landing crew. He seemed confident. He had seen plenty of situations trickier than this, and he knew that hesitation could mean trouble. He told me later that this was a lesson that he had learned early, while catching bulls in the Australian outback. With a friend driving a pickup, Winterberg would crouch in the back as they chased the animal, then leap between the horns and wrestle it to the ground. He also told me how in the Belgian military a snap decision had once saved his life. Congolese rebels had captured and strung the rest of his platoon upside down from a tree. One by one they were being executed. Winterberg managed to loosen his ropes. He chose a moment when the guards weren't looking and made a break for the jungle.

He now stood at the edge of *388*, one arm held aloft, signaling to the crane driver behind him. Extending a finger, he began to trace small circles in the air, indicating that he wanted the crane driver to spool out the line slowly. Then he balled his fist and shouted, "All stop!" The crate hung above *Abex*'s deck, the landing pad rising to meet it, pausing, then falling away again. Winterberg walked over to the crane driver and gave him instructions, thrusting his hands downward and slapping his palm with his fist. He then walked back to the edge of the deck. We all waited, breathless, mesmerized by the hanging crate and the rise and fall of the barges.

"Ready!" he shouted over his shoulder as the landing pad heaved up toward the crate again. At the zenith of its trajectory he yelled, "Go! *Now!*"

After an agonizing pause, the crane's winch screeched and the crate dropped, chasing after the falling deck. It seemed *Abex* would bottom out of the trough before they made contact. I expected a sickening smash, but at the last possible moment the crate touched down. We heard only the gentle clink of ceramics. My shoulders slackened with relief.

"All stop!" shouted Winterberg, bringing the swinging hook to a halt just above the container. A crewman leapt up and detached the cables. Our first crate was safe on deck.

The crew on *Abex* rushed over to the crate on the landing pad. Known as the "archeological" barge, *Abex* had personnel very different from those of the "dive" barge, *388*. Apart from a sprinkling of deck and mess crew, all of whom were Indonesian, the vast majority were Vietnamese — artists, draftsmen, museum staff, committee members, artifact cleaners, and guards. Everyone began crowding around the crate, hungry to see its contents.

Communication was a problem. Only Ivo and Tai were able to translate effortlessly between English and Vietnamese. Magnus Dennis, a twenty-seven-year-old Londoner who had originally been hired as the expedition photographer and videographer, had won over the Vietnamese archeologists and support crew with his own brand of full-body international communication. As a result, on top of his other duties he was put in charge of finds management. Seeing the crush around the crate, Magnus held his arms out and put his head down, blocking the approaching throng in a pantomime of protection and leaving space for Mensun and Ong to approach.

"Well, here we go," Ong said, pursing his lips. "Let's see what we've got."

The lid of the first compartment was swung open. The top layer of baskets had been carefully packed and the mud-streaked pottery seemed intact. The first few baskets were taken through the crowd to the rubber-topped washing tables. Mensun picked out the first dish with Ong looking on.

"I don't believe it," said Mensun. "Look at this . . . it's beautiful . . . half-fish, half-dragon . . ."

"The carp ascending, yes," said Ivo, who had appeared beside Mensun, his eyes glittering behind dust-speckled glasses. "According to Vietnamese legend a fish of great age can make one final leap through the rain gate—you see it here, this ring around his tail—and become a dragon. It symbolizes the overcoming of all obstacles."

Ong took the dish next and nodded, turning it in his hands. "At last," he said.

ONCE THE INITIAL excitement had passed, Magnus got his teams cleaning the artifacts and packing them into numbered desalination tanks on *Abex*'s deck, where all the artifacts would be stored until it was possible for the details of their recovery to be individually recorded. The general rule on excavations is for all finds to be tagged immediately and given a number and a reference sheet. Drawings and photographs follow, and then the piece undergoes a process of conservation. However,

with divers on the seabed almost twenty-four hours a day, recovering
thousands of pieces every shift, we had decided to delay most of the tag-
ging and recording until the excavation phase—the expensive part—
was finished. Until then the ceramics would be kept in tanks, according
to when and where they had been discovered. Mensun was uncomfortable about segregating the phases of the excavation. Even having two
barges with crews who performed such different roles worried him. He
was used to everybody being involved with and inspired by the excavation. This was the compromise he had been forced into by the expense
of saturation.

Once the crate had been unloaded, I called a boat over to shuttle me
back to *388*. As I was handing my small rucksack down to the crewman
in the inflatable, I felt a firm hand on my shoulder. I turned around and
found myself facing a soldier, with two more behind him, all looking
very serious. They pointed at my bag. Ong's decision to segregate the
crews extended to the barges. He didn't want the crews on *388* being
around the ceramics. There was no need, he reasoned, and it might be a
temptation for them. Contact between the two barges needed to be
kept to a minimum, and everyone leaving *Abex* would be searched to
make sure they were taking no artifacts with them.

Two types of soldiers were assigned to the expedition. The patrol
boat that lay off from the barges belonged to the Border Guard, which
took its orders from the Vietnamese Ministry of Defense. Based on the
military outpost on Hon La Island in the Cu Lao Cham cluster, these
men were present to protect us from external threats, such as pirates or
jealous fishermen. The second group of soldiers was represented by just
one man, the one-eyed colonel we knew as the "secret policeman." He
reported to the Ministry of the Interior, and his job was looking after
the interests of the Vietnamese government. The ministry was aware
that sizeable sums of money were at stake, both because of its experience of the Vung Tau cargo (sold six years earlier for over $7 million)
and because of what Ong was telling them. The colonel scrutinized the

daily lists of artifacts recovered and monitored all operations with a sour-faced glare.

All three of the Border Guards wore uniforms of the same pea-green color. Their faces were young, their outsized hats balanced precariously over them. The straps of their automatic rifles cut into their jackets, and half-filled shoulder pads sagged under star-laden epaulets. They searched thoroughly, first my bag and then my body. Watching the whole scene from a short distance away was the secret policeman. The colonel had another role on board: to check that the military were not abusing their authority by smuggling, or extorting from the team. I could tell from the faces of the guards that they felt the burn of his glare as much as I did.

CHAPTER 13

Indulging Mensun

BEFORE THE PROJECT began, Mensun had been concerned that some of the divers might not have the patience required to carry out archeological tasks. Having run excavations staffed by students and volunteers, or "avocationals" as they'd become known, he was used to managing a team with a mix of temperaments and skills. Some members might be less thorough than others, but they were all convinced of archeology's quest. He knew from experience that if someone found the whole purpose of archeology too abstract—as one of the commercial divers might—he would find its methods even more so, and bending them to the task would take constant supervision. The rigorous recording of all information, much of it seemingly irrelevant, drives many to distraction.

Even as a believer in the cause, I have sometimes felt my faith waver. After our failed first attempt to excavate the Hoi An wreck, I'd gone to the Cape Verde Islands off West Africa as the supervising archeologist on board the recovery vessel *Polar,* sister ship to Greenpeace's *Rainbow Warrior.* The remains of the English East Indiaman *Princess Louisa* were scattered across an exposed reef at the north of one of the islands, and mapping the distribution of trade beads and clay pipe fragments seemed an exercise in futility. Every day the Atlantic swell surged overhead with enough force to send me flying thirty feet across the seabed and back. Wedged in cracks between rocks in order to keep my position, I'd wait for the surge of a passing wave to die down, then in the two-second pause take a measurement before the suck of the next wave began. Several times the force of the flow was powerful enough to tear the regulator from my teeth, leaving me with only two pieces of torn rubber in my mouth while my air hose roared down-current. Artifacts from the wreck had collected in cracks in the reef, but several times a year storms would batter the site, reshuffling them thoroughly. Nonetheless, from chaos order can occasionally appear or an inference be drawn; however mad-seeming, method must be followed.

In 1977, George Bass and the Texas Institute of Nautical Archeology had begun working on an eleventh-century Byzantine wreck at Serçe Limani off the Turkish coast. Part of its cargo was Islamic glassware, though everything had been smashed into almost a million fragments. Initially it was assumed that the glass had broken upon sinking—the fragility of glass was, after all, why so little of it from the period survived, and thus why Bass was interested in the wreck. Having so much all in one place seemed to offer an ideal opportunity to piece some back together. He decided to divide the site into half-meter squares, recover all the fragments, and try to find segments that joined from within each grid. Nothing made sense, though; similar-looking fragments were often many grids away from one another, and the contents of some grids contained no joining fragments at all. Eventually, Bass decided to try something new. He painstakingly labeled every shard with the grid that it had come from, and then rearranged all the pieces according to the color and texture of the glass. Before long, forms entirely lost to history were emerging—from the estimated ten to twenty thousand vessels that lay smashed on the seabed, more than two hundred were reconstructed, many of them new shapes. When the team then looked at how the components of each piece had been distributed across the wreck, it became apparent that the jars and vases had been broken even before being loaded on board. The ship was carrying the smashed glass for recycling. It was good economic sense: The fragments made for good ballast, and melting existing glass required much lower temperatures than making it for the first time.

Mensun knew that using commercial divers to do archeology would be a problem, but had assumed that he could work around it by using individuals he thought more observant and cautious for his survey work and leaving the grunt work of recovery to the others. What he had not foreseen was an obstructive supervisor.

One night while off duty I went to visit Dive Control. I had been supervising the midnight-to-midday shift with Garth Sykes, while Mensun was covering the evening period with Tony Turner. As soon as I opened the door, I could tell something was wrong. Mensun was slouched, trying to look relaxed, but not quite pulling it off. Sheets of paper covered his desk, and on top of them was a logbook of detailed drawings. He was watching one of the screens intently and barely looked up when I walked in.

"One hundred and eighty-four millimeters. That's great. Now the distance from the top seam to the midpoint of the cut, directly below your last measurement?" Mensun asked, pencil held above a schematic sketch. He spoke slowly, deliberately.

Turner looked at Mensun, his hat turned around and pushed up on the back of his head like a skateboarder's. He stared vacantly for a few seconds, then lifted one of the headphones from his ear.

"Sorry, Manson. What'd you say?"

Mensun's hand sagged fractionally over the sketch. He looked over at me, exasperated, and repeated his request.

Turner's blank stare continued. Finally he lifted the headphones back onto his ear and turned back to his panel.

"Okay, Diver One. Now hear this. What these—" Turner paused, glancing slowly back through the window at Mensun and me "—aarcheologists want you to do is stick yer ruler up the top of that bit of wood and point it straight down. Yep, that's right, straight down. That's what they want. Now tell us the number on the ruler by the thingie we've been looking at."

I looked back at Mensun. His eyes were set on the page and his pencil was stabbing the drawing in front of him, stippling an area with so much force that the dots must have pushed through three sheets of paper. While Sykes accepted the twin objectives of the project without question, relaying any instructions to the diver and simplifying where necessary, Turner seemed to be deliberately confusing Mensun's instructions. In the time that I was watching he didn't pass on a single request without garbling its message, making it impossible for the diver to respond to whatever request for information Mensun had made. It was as if his brain was wired up wrong. If he heard "left," he'd say "right"; if he heard "up," he'd say "down."

Turner's tales in the mess seemed to offer a partial explanation. He had been sailing out to an oil-field job in the Persian Gulf when a missile had appeared out of nowhere, exploding only a few feet from where he stood. Hector George had also been on board but while George had escaped uninjured, Turner's eardrums were blown clean out and sixteen pieces of shrapnel peppered his body, including the "old fella," as he put it. "It was all bruised and purple, big cut across the top. I wanted to get a surgeon to fix it. I was in the corridor before the doctor came. They reckon I was trying to chase this nurse. All I was trying to do was show her the damage and get it fixed but she got all offended and screamed."

I discovered only later that Turner's manner had nothing to do with damaged or defective wiring; it had been calculated. He was a man of principle, and wouldn't stand for any "client bullshit," as he put it. "If it's common sense then I'll do it. If it's not I'll tell 'em to get fucked," he told me. Unfortunately, from Turner's point of view, Mensun and I were the client and our requests—"looking at bits of woodwork and asking too many dumb questions"—didn't seem to him like common sense.

"Right, Tony," Mensun began again in a level voice, "could you ask the diver if he can see which way the grain is going on that timber, and if there are any signs of rot or shipworm."

Turner rolled his eyes. "Course there's rot, Manson. That bloody wood's been sodden for centuries. Look, why don't we get both the boys on looking for the good stuff?"

"Tony. Finding pots is only half the game here." Mensun was angry now. "Yes, it's vital that we recover these artifacts for the sake of art history, but it's also vital that we discover their story. Without that, they are just old lumps of clay. They may be pretty, but it's their story that gives them their true value. The more we understand about who made these pots and who was buying them, the more valuable they become. And the ship is a vital part of that story. We don't know how they built these ships. No one's looked properly. Ever."

Turner blinked a couple of times, raised his eyebrows, and shrugged. "You're the boss."

"This timber we're looking at is a part of what was some of the finest engineering in the entire world. So indulge me. Let me take the time to have a good look, huh?"

Mensun's fascination with the structure of ships had first begun to flourish on the Giglio excavation. The few pieces of timber that remained on the seabed there had represented a puzzle that had gripped the young archeologist thoroughly, until he'd eventually been able to prove that the Etruscans had built the hull by using planks joined edge-to-edge, then pulled tight with a cord that threaded diagonally through the wood. It was a technique not thought to have been invented until centuries later. For over four millennia, ships had represented the pinnacle of mankind's engineering skills. Fortunes could be made with the safe delivery of just one cargo, so immense resources were poured into designing vessels to be bigger, faster, and safer. Before starting work in the South China Sea, Mensun had studied only those that had evolved in the West. Lying on the seabed below the barges was a craft that had developed in total isolation, an entirely different engineering solution to the same set of seafaring challenges.

The essential design of the junk had remained unchanged since it was first developed in the Han dynasty (220 B.C.E.–A.D. 200). The boats were soon built big. Wan Chen, a Chinese author writing in the third century, describes junks capable of carrying seven hundred people and 260 tons of cargo. However, it wasn't until 1405 that the first of the great Chinese exploration fleets set sail. In that year, sixty-two vessels departed from China. Some were dedicated to horses, others to fresh water, but in their midst were four *baochuan,* or treasure ships, over four hundred feet long and 160 feet wide, with seven masts—by far the largest wooden ships ever built. Leading the 27,800 sailors, soldiers, merchants, and diplomats of the fleet was the admiral himself, Zheng He, a seven-foot-tall eunuch from the imperial court. The fleet navigated by compass (invented in China in the eleventh century) and measured time by burning special incense sticks. Gongs, drums, flags, and carrier pigeons conveyed messages between the ships as they sailed south through the South China Sea (which they called the *fei,* or "boiling," sea, because of its churning currents) and out into the Indian Ocean. Six more expeditions followed, and eventually the imperial fleets reached India, the Maldives, the Persian Gulf, and the East African coast. They left with their cargo holds full of Chinese silks, porcelain, and other luxuries, which then were traded for foreign treasures, including a giraffe from Africa that soon became the emblem of the Ming dynasty.

For some two millennia or more, China had been convinced that it was the peak of civilization, and that it lay at the center of the world. However, Zheng He's expeditions were returning with news and ideas from other civilizations and started to be regarded with suspicion. Persuaded by conservative members of the imperial court, the Xuande emperor (1426–35) began to shut down maritime exploration and trade. By the time the Yongle emperor died in 1424, China had boasted some

3,800 vessels. Fifty years later there were only 140 warships left, barely enough to protect China from the Japanese Wako pirates that harried her coast.

When Portuguese navigator Vasco da Gama sailed around the Cape of Africa in 1497, heralding a new era of Western colonialism for the Indian Ocean and beyond, he was unaware that more than seventy years before a huge fleet of Chinese ships had traveled that same coastline in the opposite direction. The Chinese had been on a similar mission of trade and expansion, but had come far better prepared: Not only were the Chinese flagships ten times the size of the Portuguese boats (in fact, the whole of da Gama's fleet could have fit into one of the *baochuan*), but unlike the Portuguese they had come equipped with goods to exchange. When da Gama reached India he expected favorable trading terms but, as he had come without merchandise to trade, was rebuffed. Nor were native traders receptive to the Portuguese attempts to convert them to Catholicism, for they were already devout Muslims. Da Gama was furious at his double failure and eventually bombarded Calicut (today Kozhikode, located in the Indian state of Kerala), setting the tone for the next four hundred years of Dutch, English, and French imperialism.

The Portuguese called the Chinese ships that they encountered *junco*, after the Malayan word *djong*, meaning a seagoing ship. The name would later be adopted by the Dutch (*jonk*), and then the English (junk). Used to Atlantic conditions, the Portuguese were not overly impressed by the peculiar-looking vessels: They had no prow to part the water or keel to hold a course; they could not sail close to the wind and what perhaps made them even less worthy of respect in Portuguese eyes, they were not well armed. Examining their outline from above, Mensun noted how "European shipbuilders favored a fishlike form at and below the water line, whereas the Chinese sought to imitate web-footed swimming birds that have their greatest width and fullness towards the rear." The sails completed the contrast. Rather than using canvas sheets that dropped from spars, the Chinese hauled sails of woven matting up from the deck, stiffening batons protruding like spines. They lasted well, for their rigidity prevented shakes, the flapping that eventually splits canvas sails; and when bad weather threatened, rather than sending men aloft to reef and stow the canvas, the junk's sails could simply be dropped to the deck. Their disadvantage reflected the relationship that the Chinese had with the seas: Sails could not be spread in a hurry when a threat arose from an enemy vessel. However, this rarely happened, and the Chinese traders timed their trips with seasonal monsoons that could be relied on to blow from behind, so were under less pressure to sail close to the wind. That did not mean they were oblivious to the advantages of the rigging of the Europeans, who "make sails like spider's webs, which can be set at any angle to catch the wind," in the words of one contemporary Chinese diarist.

Other details of the junk spoke volumes about Chinese philosophy. On a European vessel, every element of the structure was polished and oiled as a matter of pride; the Chinese saw no point in sanding or even smoothing areas that would not normally be noticed. Masts were not ramrod-straight or even circular in cross-section; the curved forms found in nature were thought to be stronger, so masts simply consisted of seasoned trees whose bark had been stripped.

The unchanging template of junk design had been preserved by Chinese law, which would penalize ships that deviated from the old and established rule by charging port duties as if they were foreign vessels. Abroad, there were no such constraints. Other Asian shipwrights, and those who had fled China after the ban on building oceangoing vessels, began to incorporate aspects of Western ships that they saw as superior. Junks began to take on nontraditional elements, such as keels (known as "dragon spines" that helped captains avoid rocks), sternposts, and swim heads.

As Mensun had told the dive crew in his briefing, information about these remarkable ships—often claimed to be the most successful design in maritime history—is sparse compared to academic knowledge of Western vessels. The archives of the Spanish, Dutch, and English shipping companies and navies describe in detail how their craft were created from the hundreds of trees that it took to build each one. Excavations of well-documented ships such as the *Mary Rose* yield invaluable details that were never recorded at the time, such as how ordinary seamen lived and how vessels were repaired. Only the earliest European ships of exploration—those of the Portuguese—lacked accurate descriptions; most of Portugal's archives were lost in the fire that consumed Lisbon after the great earthquake in 1753. The few underwater excavations of Portuguese ships, such as the *Santo António* (1589) off the Seychelles and the *Santiago* (1585) off India, have since shown how heavily timbered the Portuguese caravels and *nãos* were, implying still-ample supplies of hardwood and a degree of cautious overengineering.

Chinese shipwrights didn't make plans from which to work, and what information was recorded was almost entirely lost, destroyed not by accident but by intention. As a result of the xenophobic paranoia surrounding the fifteenth-century Ming ban, all records of exploratory endeavors, including how the ships had been built, were burned, while fighting at the end of the Ming dynasty destroyed all but ten thousand of the several million documents once held in the central government archives in Beijing and Nanjing. Records from subsequent dynasties were torched during the "Cultural Revolution," when in a single decade, from 1966 to 1976, Chairman Mao's enforcers successfully erased five thousand years of China's imperial history.

"ERR . . . MANSON?" Turner's voice sounded a little wary, given Mensun's last outburst. "Wanna take a look at Diver Two's camera? He says he's got something a little bit different."

Both Mensun and I turned to the screen. For the last half hour it had shown little more than swirling brown silt, the diver shrouded in mud as he removed a layer of pots from a grid. Now the swirling had slowed, and a white patch began to appear, like a break in the clouds. A figure appeared, cradled in the diver's hand. Mud streamed from it, making it look as though it were wreathed in smoke.

"Whoa!" exclaimed Mensun. His whole body shifted upright. "Hold it right there!" His eyes were fixed on the monitor. "Okay, bring it closer, higher. Left a bit . . ." This time, Turner relayed the message to the diver word for word.

The ceramic figure now occupied the center of the screen. It was a statuette of a robed figure kneeling, an urn in front of him. A wise-looking face stared out at us from beneath a tall hat. The front of the urn was chipped, as was the top of the hat. Otherwise the statuette was intact. Mensun shook his head as the figure was rotated slowly for the camera.

"This is fantastic! A mandarin! I was scared, I have to admit . . . ," Mensun began, then caught himself. "Tony, can he tell me where exactly in the grid the statue was found? In what orientation? What it was packed in? What it was near?"

There was no hesitation now as Turner transmitted the questions to the diver, his eyes alive and enthusiastic. "What d'you think he's worth, Manson?"

Mensun laughed. "I don't know. Lots. Fifty, maybe one hundred thousand dollars. I just don't know. Let's get him in the specials basket."

Turner was evidently impressed. "Right-o." The "specials basket" was a fast-track route to the surface. Padded with foam and full of toweling that was used to wrap and pack valuable pieces, the basket could be hauled to the surface independently of the main crate and brought straight into the archeological section of Dive Control. These were the only artifacts to touch the deck of 388, Ong and the soldiers having consented to make an exception for them.

Fifteen minutes later, the basket with the statuette was sitting on the floor in front of us. I unwrapped the sodden toweling carefully. From it emerged the mandarin, as if materializing out of the monitor. My reaction was quite unlike any I'd experienced from touching an object before. The top of my arms and the back of my neck began to tingle. Until that moment the Hoi An shipwreck and its ceramics had been abstract for me, separated by the screen on which I had been watching everything unfold. Somehow the few pots that I had seen, though beautiful, had simply been remnants of a historical period. Looking at the mandarin's expression, however, I now felt as if I were being let in on a great secret. All the troubles we'd had getting to this point seemed to melt away under his gaze. Grinning, I looked up and saw Turner smiling beatifically across the desk. Mensun's eyes were twinkling as well as he carefully took the mandarin from my hands and ran his fingers reverently over its contours.

"It's uncanny," he said, shaking his head. "This looks like it could even be from the same mold as the one I saw in Oxford. That was at least five years ago. Those looters must have raked this wreck for all this time. God, I'm surprised that any of this stuff has survived. But it has. And look at it. Look at him. He says it all. "

ONE THOUSAND MILES to the south, other discoveries were being made. Mike Hatcher had not been idle since paying *Abex* a visit at the Pandan Road dockyards the previous year. As we worked off the Vietnamese coast he was scouring the shorelines of Indonesia in his pirate ship *Restless M,* looking for a wreck that would yield him treasures on a par with the mandarin statuette. Using a combination of side-scan sonar, magnetometers, and diver swim-searches, he'd been concentrating on areas that looked likely on the maritime charts. Outlying rocks, reefs that lay just below the surface, and those in the path of prevailing winds were always good bets, but Hatcher's search techniques were more refined than this by now. He knew that in the centuries when the porcelain trade was booming, one of the major sea routes was between the Chinese ports and the Dutch entrepôt of Batavia (present-day Jakarta). Tracing a straight line between the two, south from the Gulf of Tonkin and the porcelain-producing lands of southern China, ships would pass the coast of Vietnam, across the mouth of the Gulf of Thailand, and into the funnel created by the Malaysian peninsula and Sumatra on one side, Borneo on the other. Toward the bottom of this funnel, a few degrees south of the equator, any ship captain aiming for Batavia would have to make a choice: Go through a narrow western passage between Banka Island and Sumatra, or between Banka Island and Belitung Island—the Gaspar Straits. The straighter line is that which cuts through the Gaspar Straits, and perhaps that's why Hatcher was searching there, not realizing that most trading ships had taken the western passage because it was the one that had been more accurately surveyed. Whatever the reason, it was a fortunate decision for the treasure-hunter.

In January 1822, a large junk named the *Tek Sing* (or True Star) had set sail from Amoy (modern-day Xiamen), bound for Batavia. She was heavily laden with blue-and-white porcelain from Fujian province together with almost two thousand crew and passengers, mostly emigrants leaving to work in the sugarcane fields of Java. Coming down through the funnel toward his destination, the captain decided to try his luck going through the Gaspar Straits. Perhaps he was wary of the pirates that lurked in the western passage, or the winds were wrong; perhaps he wanted to make up time. Whatever the reason, it was a mistake. When his ship hit rocks that weren't marked on his charts, she went down fast. Only 198 out of the two thousand people on board survived.

Sometime early in 1999, running search lines near a reef some twelve miles north of Gaspar Island, Hatcher's magnetometer registered a small metallic anomaly on the seabed. When Hatcher and his divers went to investigate, they came across the concreted remains of three iron rings, each over three feet in diameter. Straps used to strengthen an enormous mast perhaps one hundred feet tall, the rings led back to a prominent rise in the seabed at least 150 feet long, thirty feet wide, and six feet tall. All Hatcher's instincts would have screamed that this was a shipwreck, and that within that mound would be either ballast or cargo. Digging into the sand with his hand, Hatcher pulled out a dish, then another. Porcelain. And, though he was no expert, it looked Chinese.

Back on the surface there was celebration, but no one on board yet knew the identity of the ship that they'd found. The dish and the location were a good starting point, however, and Hatcher hired Nigel Pickford, a renowned shipwreck researcher, to start investigating. Armed with photos of the dish, or perhaps even the dish itself, Pickford approached a porcelain specialist. The diagnosis didn't take long. Manufactured in the workshops of Dehua in Fujian, the piece was made for sale in the Southeast Asian markets. It was from the Qing dynasty, and dated from between 1815 and 1830. Given where the ship had sunk and knowing where she'd come from, her destination had to have been Batavia, so Pickford's next step was to start searching the archives for all big ships that sank en route from China for Batavia between those dates. It didn't take long to find; given the loss of life, the *Tek Sing* disaster was well documented at the time, a tragedy known to have taken place somewhere between Borneo and Sumatra. The only curious thing was the absence of any mention of the ceramic cargo. Still, the dates and the approximate location fitted, and the story was a dramatic one. Hatcher seized on it, immediately adopting the tag line "the Chinese *Titanic*." The porcelain was not from a particularly interesting or rare era, experts warned, but the size of the cargo and the story of the *Tek Sing* were an irresistible combination that Hatcher knew he could sell.

Legally, Hatcher should have had a search license before even dropping his magnetometer into the water, but in reality treasure-hunters tend to do the searching first, then secure the license after having found something worth looking for. The process costs time and money, after all. Whether Hatcher waited for his license before starting recovery is uncertain, but through middlemen, an Indonesian company was established that was soon awarded one. By mid-May Hatcher's team had begun work in earnest, the divers filling plastic bins with porcelain as fast as they could. When one bin was finished they'd send it up on ropes while they started stuffing the next.

At the other end of the South China Sea, Ong and Mensun were blissfully unaware of Hatcher's discovery and the effects it would have on their ambitions.

CHAPTER 14

The Dragon Strikes

ON APRIL 27, I woke up early. The last of the evening light was dying in heavy air. My shift wasn't due to start until midnight, but I'd been awakened by the barge's lazy rolling. It was the first time I'd felt *388* moved by the sea. A long cross-swell was running, the faces of the waves roughened by a purposeful wind that blew to our bow. The bell was off the chamber, still underwater, the crane stowed. It took me a few seconds to realize what was wrong with the view. *Abex* was gone. I walked to the port side of the upper deck, then to the starboard. Nothing.

I found Dilip outside the barge office on the deck. He looked distracted. "She's been taken back to Cu Lao Cham," he said, without my having to ask. "She can sit there in the bay instead of here. The cross-swell was making all the Vietnamese sick."

"Is there weather coming?"

"Just safer this way. For the time being we'll transport the crates using *Ena Supply*. They can get offloaded in the bay. It'll work fine. Don't worry."

I wasn't worried, but I was surprised. This was a substantial change of plan. The barges were meant to remain close to one another so the crane on *388* could land the crates of ceramics on *Abex*'s stern. Cu Lao Cham Island was more than an hour away. *Ena Supply* was a big tug for carrying a five-ton crate. Her 2,500-horsepower engines would burn a lot of fuel, and she might have to make the trip two or three times a day. Besides, neither *Ena Supply* nor *Abex* had a crane to transfer the crate between them. And with the dive barge and the archeological barge so far apart, how would the teams be able to communicate properly? A dozen questions and objections flew into my head.

But Dilip wasn't in the mood to talk. The satellite phone began to ring and he disappeared into the office. I stood for a moment, lost in thought, watching the sea roll past, when I heard a voice above me.

"Ominous, that."

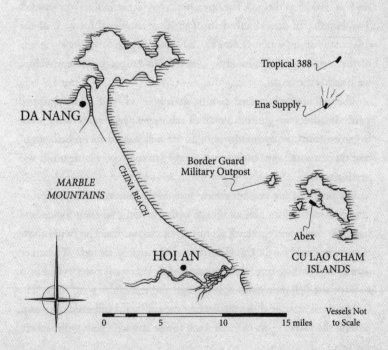

I looked up. Sykes was propped up against the rail outside his cabin, a can of soda in his hands.

"What do you mean?"

"Over there. Look." He tilted his head back, pointing with his aquiline nose. A loose line of five or six boats was moving slowly across the horizon. Fishermen. Behind them I saw more, perhaps the same number again, strung out along my field of vision. They were all headed toward Cu Lao Cham. "If the fishermen start going in, you'd best be thinking about it too," Sykes said, taking another sip from his can.

GAZING OVER THE surface of the water, Sykes and I were oblivious to the fear that was mounting 220 feet below. Roland Dashu was sitting on the bare metal seat in the dive bell, holding on tight. The water beneath his feet boiled up through the open hatchway as the bell fell downward. Two bundles of hoses snaked through it, an umbilical for each of the divers still in the water. Dashu braced himself against the opposite side of the capsule with his hand squeezed amid gauges, pipes, and valves, waiting for the recoil.

The bell lurched upward, pulling sharply to one side. Dashu gripped hard, dreading the gunshot crack of cables snapping apart. The water began to churn in the moon-pool as the bell started its upward surge, and then erupted into turbulence as it began to drop once again. It was getting worse.

He clenched his jaw, breathing fast. How long was the supervisor going to wait before pulling up the bell? He had told them twice how hard the cables were jerking against their mountings. The barge above must be rolling heavily. Dangling from the port side, the bell was yanked upward when the barge tipped to starboard. When the barge rolled back to port, the bell didn't sink fast enough and the cables went slack. The bell was still floating downward when the starboard roll caught it again, pulling the cables—his elevator back to the surface—taut with sickening force.

Dashu swore, beads of sweat joining the condensation that had dripped onto his face. He was unaware that just a few feet from his head, one of the U-bolts holding the bell to its guide wires was beginning to bend under the strain.

BACK ON THE SURFACE, Spanner squinted up at the bell winch, watching the cables slacken and then snap rigid. He pushed his glasses back up the bridge of his nose and stared out to sea. How bad things were allowed to get before pulling up the divers or running for shelter was always a bone of contention. The client—Ong, in this case—always wanted to stay and the dive contractor always wanted to leave. Minimum swell heights and wind strengths had been negotiated and laid out in a contract before the project began, but in the end it came down to the judgment of those in charge on site. Spanner looked over at Turner and gave him the nod. There was no point keeping the bell down any longer; the weather wasn't going to improve.

SIX HOURS LATER Dilip was in the barge office, his finger running over the thin thermal paper of a Weatherfax printout. The automatic transmissions from weather stations dotted around the world could be received by anyone with the right equipment, and showed a situation report for the local area.

"It doesn't mean anything. So Taiwan doesn't transmit. We change station and we get the information from somewhere else. And now we have the report," Winterberg declared, looming over the desk, leaning on his fists.

"Could be," said Sykes quietly. "But it seems strange that a tropical depression disappears just like that." He put his hand on a small pile of Weatherfaxes from the two previous days. They showed a deepening depression approaching over Taiwan, getting closer to the barges. The next report from Taiwan didn't come in, so Dilip changed frequency to another weather station in the Philippines, but the low-pressure system it showed was markedly less intense.

"Okay. Maybe she's still out there, but what do you want to do, run from this?" Winterberg gestured at the window, on which raindrops were scurrying diagonally downward. "It's a breeze and some rain. Sure, we can't dive, but we can sit this out no problem. Right, Dilip?"

Dilip stared at the page for a few long seconds. It was a difficult call. Static from an electrical storm had caught him out last year, and this time the stakes were even higher. Picking up anchors and heading for shelter now would mean losing at least two days getting back on site. Over the phone, Ong had requested him to try to ride it out.

In theory the contract would call the shots. If a named tropical storm came within a certain distance, the dive contractor's agreement with Ong dictated that the barge must seek shelter. According to Dilip, this alert radius had initially been five hundred kilometers (three hundred miles), but Ong wanted it changed to two hundred kilometers (125 miles). With the islands of Cu Lao Cham so close by it would be a quick run to safety, Ong had argued. But with nothing showing on the Weather-fax, it didn't matter what the alert radius was.

"Guys, we don't even have a tropical depression, let alone a storm," Dilip said. "I've checked with Josafat on *Ena Supply*, and he can't raise Taiwan either. His report from the Philippines reads the same as ours. Nothing. All we can do is keep checking forecasts and hold position. In three hours we'll have another look."

"I don't need to tell you this, Dilip, but I'm going to say it anyway," Sykes said, settling back against the wall, his arms folded across his chest. "We're much better off cutting and running before anything gets seri-ous. Easy out, easy in. I've been watching fishing boats running for shel-ter all afternoon. Right now we can still get our anchors up nice and easy, we can retreat nice and dignified and be back up and running within a day of the weather clearing. Leave it too late—"

"Garth, I know the implications," Dilip replied. And he did. If the weather did get worse and they had to run, they'd have to drop their an-chors—it would be too rough to recover them. That would mean much longer delays. Dilip stared out of the rain-streaked window, looking for some clue as to what the weather would do next.

TWELVE HOURS LATER, peering through the scratched, two-inch-thick glass of the main chamber's porthole, Dashu could barely see the rain lashing past the deck lights. When at last the bell had been returned to the system on the barge, he had felt as though a weight had come off his chest. The continual jerking of the guide wires had jangled his nerves badly. While in the bell he had been desperate to get back to the surface, but when he had climbed through the narrow trunking into the damp sphere of the transfer lock his limbs became heavy with dread. There was no way out of these metal walls. If anything went wrong, that was it.

The feeling got worse. The interior of the chamber was disorienting and his sense of balance had been thrown off by high pressure and helium. Even so, he was sure he could feel the barge wallowing, its movement unnatural and heavy. Were they taking on water? Had it started already? Twice now he had asked Robbie what was going on outside, how the weather looked, and been palmed off with platitudes: "Yeah, it's all okay out here. Bit of rain, bit of a swell. We'll be right." Dashu knew Robbie wouldn't tell them if things were bad. There was no point in disturbing the divers, he'd say. There was nothing they could do. Dashu kept a constant eye on the porthole. There was a saturation diver's code of honor to get around a protective life-support technician: If anything serious was happening, one of the divers outside on deck would put a note against the glass to let the divers in the chamber know. So far there had been no note. Maybe the others were all too busy trying to save the barge.

Dashu looked over at Kerr and George. They were flat on their backs in their bunks, their high-pitched snores shifting in and out of synch. How could they sleep? The tubular steel saturation chambers amplified and distorted all outside noises, which joined the whirr of the carbon dioxide scrubbers and the constant hiss of gas. The valves and gauges

crowded the walls, indicator needles trembling in the precarious, unstable environment. Dashu the medical technician recognized in himself the onset of claustrophobic panic. He wanted to escape through that
porthole, get out on deck even for a minute. But he knew he would die
in seconds. There was no escape.

The chamber shrank in his eyes, feverishly warping in size. His mind
raced. Escape. The HRC. The Hyperbaric Rescue Craft. Since the
DB-29 disaster, every saturation vessel was supposed to have a pressurized escape vehicle. The one on this system was token at best, a capsule
that had no propulsion—hence its being called a craft, not a vehicle. It
doubled as the other team's sleeping quarters, and was bolted onto the
main system. Bolted. They weren't explosive bolts, either. What if they
had to get it off in a hurry? Besides, how could they even get it in the
water in weather like this? There was no launch derrick, and they
wouldn't be able to use the crane. There was no extra flotation on the
craft, either. It hadn't arrived in time. The HRC was designed to float
without it, but had anyone tested it? Not while he had been looking.

None of this had crossed his mind when they were fitting the system
out in Singapore; now it was all he could think about. No one had ever
used an HRC—whenever saturation barges had gone down, they had
not been fitted. In the North Sea, regulations now required them to
be self-propelled, carry a week's supply of on-board gas, and be self-
contained for decompression. They also had to accommodate two un-
pressurized life-support technicians who could monitor the atmosphere.
But this was the South China Sea. This HRC had nothing. An Emergency Position Indicating Radio Beacon (EPIRB), a few cylinders of
gas—maybe a day's worth. There was no Environmental Control Unit,
either. Helium and high pressure combined to make the body sensitive
to temperature. Even if the HRC did float, without the means to regulate the internal temperature they'd all die of exposure if the weather
stayed bad, or fry if it got hot. Dashu's breathing was getting faster. He
couldn't shut his eyes at all.

———

"JESUS CHRIST! Where'd that come from?!" Dilip shouted, staring at the rain-spattered Weatherfax that Winterberg had passed to him. We had been wondering at the ferocity of the unforecast wind. The howling had changed pitch, and now it groaned through the superstructure. The slapping of the awning over the stern had become a machine-gun rattle.

"Says on the report that Typhoon Leo knocked out the weather station in Taiwan before it could broadcast. It's back up now. That's how we've got this," Winterberg shouted back, trying to be heard over the noise. "But look—I think it's going to pass north of us. Our anchors are holding well. It's safer for us to stay here than try to move!"

Dilip considered the fax. Winterberg was framed in the doorway, getting buffeted by the wind even with his legs braced wide. "It's well inside our alert radius!" announced Dilip finally. Winterberg opened his mouth, but Dilip cut him short. "There's divers in the pot. We've no choice. Get the towing bridle ready. Drop three anchors. I'm calling in *Ena Supply*!"

The barge was Winterberg's responsibility, but Dilip was the boss. The windswept decks were suddenly full of crews in storm gear. Winterberg marshaled his men, preparing to drop all but one of the anchors. That would keep the barge more stable against the wind and sea, helping the tug to maneuver close enough to take the towing bridle.

The dive teams began an immediate emergency decompression of the chambers. This was standard procedure when trouble threatened, designed to give those in saturation the best possible chance of survival. It was a token precaution; in all likelihood a storm would either have claimed its victims or blown itself out well before the men would be ready to emerge.

The wind was now blowing at Force 8 strength. *Ena Supply* bucked heavily on the swell, rearing up and crashing into troughs thirty feet below. Gradually she approached *388*, her captain showing fine control of his vessel. As the gap closed, *338*'s bosun threw across a messenger line. Spread along the gunwale, the tug's crew caught the line's balled end and reeled it in, then started hauling across the heavy rope towing bridle. Once it was secured, the tug surged forward, pushing upwind for what seemed an age. She had disappeared into the swirling rain for some time when the towing bridle finally pulled tight.

Winterberg had cleared everyone well back. The ropes, as thick as a man's leg, snapped taut. They stretched from bollards, the heavy-duty iron posts that were fixed at each corner of the bow, joining into a single cable that ran to the tug. With the strain now taken off the last remaining anchor, its cable flapped loosely. A crewman lashed a buoy to the cable beyond the rollers to allow it to be easily recovered later, then what was left on the winch was released overboard. We were free at last, dangling behind the strong engines of *Ena Supply*. She began to pull, making for the shelter of the Cu Lao Cham islands, where we'd join *Abex*.

The tug was more than six hundred feet away and the sound of her engines was distant. The only other indication we had of her presence was the surging pull of the towline. When the barge scooted forward on the swell, the cable dipped into the water, sagging. Then, as though struck by lightning, it would suddenly thump straight, tearing up through the waves, the water hissing from its fibers before beginning to creak. The tension slackened, the hawser dipped into the waves, and the cycle started again.

The towline survived for less than forty minutes. With a terrifying crack it snapped two hundred feet ahead of us, rearing up like an angry cobra before collapsing into the sea. The scene hung for a long second, as though the snap had frozen the moment in a sonic flash. Then— chaos. In an instant, the barge wheeled around in the wind. Winterberg's radio began to squawk as he bellowed at the crew.

Tropical 388 was now at the mercy of the storm, and with the entire superstructure acting as a sail, was being blown downwind fast. It was essential to get another towline attached to the tug immediately. The outer reefs and cliffs of the islands were not far away. A second towing bridle had been prepared but was still attached to the stern for use in case the weather had come from the other direction. The storm was worsening by the minute and time running out. Winterberg decided to leave the bridle where it was and pick it up from behind.

Ena Supply circled back around, lights cutting through swirls of rain so thick they looked like mist. Passing a messenger line from our deck to hers while drifting would be a challenge. At first the tug tried to approach forward, her crew lined along her bow, but throwing the line against the wind up to the high, oceangoing prow proved impossible. Her captain, Josafat Padoma, pulled away, concerned that *388*'s bow might swing around too fast for him to get his tug out of the way in time. Getting close enough to take the line would require all his skill. In these seas, any contact between the vessels would be catastrophic.

Unable to help, I climbed onto the upper railings on the back deck and watched as *Ena Supply* circled again, looking for a way to approach stern-first instead. Drifting gaps in the clouds revealed dark shapes— the islands were rapidly approaching. Padoma backed *Ena Supply* toward our port stern, using the lee of the barge to avoid the wind and retain control while giving Winterberg a chance to throw the line downwind. Almost on a level with the tug's bridge, I could see the tension in Padoma's movements as he stood braced by his controls, pushing the low stern deck of his vessel ever closer to the drifting barge. Winterberg and four crewmen were clustered on the corner of *388*. He now had the coil of rope and the knot. He had tied a big steel shackle to the rope as well, adding weight to help the line across.

Crewmen in bright yellow storm gear clutched the rails that lined the open back deck of the tug. At every pitch and fall of the vessel they were swamped with waves that forced them to hold on tighter still. Fifty feet separated the vessels now, and the difference in the motion of the two was terrifying—one moment the tug was far below us, the next far above. Winterberg chose his moment and hoisted the line. It arced high in the air and looked promising, until it was swatted down by a gust. He tried again, to no avail. He grabbed the radio from the bosun standing by his side.

"*ENA SUPPLY! ENA SUPPLY!* CLOSER! IT'S NO GOOD! YOU'LL HAVE TO COME CLOSER!" he shouted. Passing the radio back, he tore his sodden shirt open and pulled it off his body, swinging his throwing arm to loosen it up.

Sykes was yelling in my ear. "This is no good . . . no good at all!" He was one of a few who had come up to watch their fate while keeping out of the way of the deck crew. "We should have a gun to shoot that rope. This is bloody dangerous. We've got guys inside here. If we so much as touch that tug we'll be ripped open like a tin can. We're heavy enough as it is. Flood a compartment and that's it." Sykes's face was stretched, eyes narrowed. I'd never seen him like this before. Looking beyond him at the other dive crew, I saw they were all focused intently on what was happening below, their faces grim.

Blown by the storm, we were bearing down upon the back deck of the tug, the two vessels now horrifyingly close. Winterberg threw the line again, then again. Once there was a metallic clang as the shackle hit the back deck and a crewman lunged for it, but a breaking wave got there first, sweeping the rope away and nearly taking the crewman with it. Winterberg hauled the line in as quickly as he could, trying to coil it carefully so that it would not catch as he threw. Padoma edged his tug closer, while beyond his bow the islands appeared with increasing frequency.

At last Winterberg judged the tug close enough. He swung the shackle twice, three times, and then heaved it across. It flew diagonally across the swamped back deck and clattered against the metal. Three yellow jackets leapt for it, skidding unattached across the exposed steel. *Tropical 388* reared up as a wave—the largest of any so far—heaved beneath us. As we rose the tug sank, sucked into the rogue wave's trough. My heart clenched as I watched the tug's stern disappear beneath the corner of the barge, one crewman sliding backward across the deck, clutching the messenger line while the others looked up in terror at the bulk of *388* rearing above them.

For an agonizing second we hung in the air before beginning a sickening drop downward. The corner of the barge plunged down as the tug's stern scythed upward, its corner passing less than three feet from the thin skin of our hull. As the tug heaved back upward on the next wave I could see the sliding crewman lying prostrate on the edge of the stern deck, clutching a crewmate with one hand and the messenger rope with the other. He had obviously only just stopped himself from going overboard.

"GO! GO! GO!" Winterberg was shouting into his radio. Padoma had already seen that his crewman had caught the line, and the water between the vessels was churning white from *Ena Supply*'s propellers as they clawed at the sea, trying to generate some space between us.

With agonizing slowness the tug hauled 388's stern to face the weather, waves crashing awkwardly over the blunt end. Looking back over the bow, I could see breakers at the base of the looming dark islands. They had been that close. They faded slowly as the tug pulled us away. Gathered in clusters to digest the last few hours, the others looked as relieved as I felt, and by late afternoon we had joined *Abex* and the fishing fleet in the protective embrace of Cu Lao Cham's bay.

CHAPTER 15

Hope and Pray

I STOOD AT THE stern of 388 with a steaming cup of Vietnamese tea, watching as eddies of wind whipped around the islands and into the bay, making ripples flee in shoals across the water. Behind the green hillsides of Cu Lao Cham, the sky remained heavy and dark.

"Mmm. Very early for typhoon! Maybe the sea angry!" said Nguyen Thuong Hy, his eyes wide and shining with laughter above his Himalayan cheekbones. The Vietnamese artist wore an extra-large expedition T-shirt that covered him like a sheet and delicate hands danced on the end of stick-thin arms as he spoke. A representative of the Quang Nam museum, Mr. Hy had followed the Hoi An shipwreck project since the first surveys. He could have overseen the project for the museum, but had been humbled by the thought of so many unprecedented artworks. Instead he volunteered his services as a draftsman, for though his own art was modern in style he was fascinated by what his ancestors had painted.

"Is that what the fishermen say?" I asked.

"They angry too, maybe," he said, suddenly serious.

"Angry at us?" I was confused.

"Maybe our prayers not well received. Maybe we take things we shouldn't."

I thought of the mandarin, now incarcerated in a desalination tank on board *Abex*. "And what about all the things that were taken before we came?"

"Ah!" Mr. Hy smiled again. "Different! The fisherman didn't go inside the water and take, he take only what he was given!"

LESS THAN one hundred feet away from us, Dashu lay rigid on his narrow bunk inside the saturation system. Eyes wide open to the darkness, he flinched as the carbon dioxide scrubber next to him suddenly whirred into life. While high-pitched snores still came from the other bunks, his own breath came only in short little gasps. Gripped by claustrophobia, he felt as though the curved, damp walls of the chamber were closing in. The sickening motion of the storm had ceased, but his thumbs remained gripped in his fists and, for the hundredth time that night, he fought the panic rising in him. The only thing keeping his hands from scrabbling at the hatch was the slow deafness that built up in his ears, a deafness that disappeared when he swallowed, a sign that the chambers were slowly returning to atmospheric pressure. They were decompressing. In a little over two days they would all be out, or so he thought.

LOOKING OUT at the sea beyond the saturation system, Dilip cut a lonely figure standing on the walkway outside his cabin on the upper accommodation deck. His phone was at his ear, as it had been ever since we'd been back in the bay. His face had become increasingly drawn. He told me that Ong was furious that the anchors had been dropped once again and was calling at least once an hour for weather updates. Dilip felt that he was being personally blamed for the storm.

Dilip hung up and leaned back against the railing, shaking his head. What's more, he said, Ong had decided—from wherever in the world he was right then—that the weather must be clearing, and that the team must be ready to go the minute there was an opportunity. Dilip had tried to explain that canceling the emergency decompression of the saturation system would not save time. If the storm carried on all week, it would be a massive waste of money to keep the divers inside. Ong was insistent, and in the end Dilip gave in and passed on the message to Robinson in Dive Control. The decompression would be cancelled. Though it was frustrating to have his advice ignored, it wasn't his problem.

Ong apparently wanted him to go straight back out with the barges too, however, and that *was* his problem. He had explained that not only was there no point, it was impossible. First they had to retrieve the anchors and spool the cables back onto 388's winches, and the wind outside the bay was still blowing Force 8, at least. It was madness. You couldn't fight weather systems with pure bravado. With a dynamic-positioning vessel, maybe, Dilip said bitterly.

"I'm going out to the site first thing tomorrow," he told Mensun and me when he came down. "To try to pick up the anchors with *Ena Supply*."

"That's ridiculous!" said Mensun. "There's no way you can work out there. Not one of the fishing boats has even thought of moving. Look at them!"

"I know. Believe me, I know. But the boss doesn't believe me that the sea's too bad. He's insistent, says that we *must* get back on site. We've blown the schedule already. We have no slack left, apparently. We're losing fifty thousand dollars a day sitting here, so he wants me to go and sit out there instead. Fine by me."

"Going out there now is only going to make things worse. The tail of a typhoon is still thrashing around out there. Ong's not here, he can't know what it's like."

Dilip shrugged. He didn't mind going out on the tug. Each anchor would have to be snagged by its pendant buoy and hauled on deck along with its chain and cables. He had to wait for breaks in the weather, then see if he could snatch the anchors one by one. It beat sitting in the bay getting abused over the phone.

At first light the next morning *Ena Supply* rumbled out to sea, watched by every eye in the bay.

CHAPTER 16

Fresh Blood

ROBBIE ROBINSON leaned against the open door of Dive Control and pushed his cap—marked ROBBIE in big gold capitals on the front—farther back on his head. He watched *Ena Supply* round the headland and start to buck and rear in the unprotected sea beyond. When the tug disappeared, he looked up at the heavy gray sky and a sigh whistled through his teeth. So Ong and Dilip were going to keep the divers in saturation after all. It didn't make sense to him, but he was too long in the tooth to question it. At least the guys would be happy, getting full saturation wages for doing nothing except reading paperbacks and motorbike magazines.

As the life-support technician, Robbie was in charge of more than just the divers' physical well-being; he also had to keep them sane. Being cooped up in a confined space for weeks on end was bad enough. Add to that the dangerous nature of the work and the kind of people you were likely to work with, and the potential for losing one's mind was high.

Though the supervisors and support crews were only feet away from the chambers where the divers ate and slept, those inside grew heavily reliant upon each other. If anything went wrong, if one of them had a stroke, or got appendicitis, or was injured, they were days away from help from anybody else. Feeling responsible for themselves and self-sufficient, they begin to resent outside authority. This syndrome is known as siege mentality, or "us versus them," and develops wherever teams are sent to work in remote and extreme conditions: space, the polar regions, or the deep sea. Facing tough conditions and having to do the work themselves, they see those back at base or on the surface as soft, weak, and ignorant of the team's situation. The syndrome even caused a full-blown strike on Skylab 4, a U.S. space mission launched in 1973. The crew was faced with a punishing work schedule, but when they complained, the ground crews told them they weren't working hard enough. Having to eat an experimental diet that excluded salt and other condiments, their morale sank to the point that they put down their tools and refused to continue until their workload was reduced.

The perceived isolation made Robbie's job more difficult, because the divers stopped letting him know if anything was wrong. He had to watch and listen carefully. Dashu had sounded a little jumpy the last day or two, but that was only natural after a few years off the job. It was easy to idealize saturation work in your mind when you hadn't done it for a while; you remembered the adventure and the money and forgot how hard it was. Robbie wasn't expecting any trouble yet, however. The guys had been inside for only a week, and though they'd been doing twelve-hour shifts in the water, they'd had a couple of days off and were about to get a whole lot more.

Robbie had been a saturation diver in the early days and a supervisor for decades, so he knew all about the stresses and strains that could build up inside. He had begun his diving career on these very shores for the Australian navy, sneaking up estuaries to clear mines before troop landings during the Vietnam War. That was thirty years before, when the

Australian Special Forces were showing the Americans how a war in these parts should be fought. They avoided big confrontations and fought the Vietcong on level terms as guerrillas, sneaking through the jungle in small patrols, never following paths and carrying supplies that would enable them to lie low for days. They fired less ammunition than the U.S. Special Forces and made sure they were popular with the villagers who lived where they patrolled. Though the American generals thought their tactics too detailed and that the war would take too long to fight this way, they couldn't argue with the antipodean mission statistics. The combined special forces of New Zealand and Australia achieved kill ratios of 500:1, higher than almost any other force at any time, anywhere. As a result, men like Robbie were taken seriously.

The commercial diving Robbie had done since the war had been gritty work. He'd stuck at it mainly for the money. He loved his kids and wanted to give them a good start in life, so it was with mixed feelings that Robbie watched his son Chad walk up to Dive Control, carrying a tray of food for the divers. It was good to have him on board, but he didn't want Chad to end up like him.

"Ah. That my breakfast, is it?" Robbie called out of the open door of Dive Control.

"I'd say yes but that Maori'd eat me if I gave it to you, I reckon," Chad replied with a grin. "We set to lock it in?"

"Hold on a second, let me just check what's up inside," Robbie said and turned to the panel. The most dangerous moment was when anything was transferred into or out of the chambers—whether divers, equipment, or food. Get it wrong and the whole chamber could vent its highly pressurized gas in under a second. The med-lock, a small air lock into the main chamber designed for getting medicines inside, was a notorious weak point because it was used so often—at least four times daily, to get food to the divers. Some divers move into the transfer-under-pressure compartment with the hatches shut every time the med-lock is opened, just in case.

Robbie checked that the inner hatch was sealed, then waved Chad ahead with the tray. "Yup, you're okay there, son. We'll let 'em eat, then let them have the good news."

DASHU PUSHED the pieces of egg around the plate in front of him. They were tepid and tasteless. Helium leached out any heat remaining in the food after it came from the kitchens and through the med-lock. It also took away his sense of taste and smell. His skin crawled as he watched Kerr and George demolish their breakfasts. How could they be hungry? Dashu was trying hard to hide his claustrophobia. His teammates weren't the most understanding of companions. Just his luck to be stuck inside with two ex-Special Forces madmen. Initially the lack of safety mechanisms had set him off. Now everything about being inside these chambers made him desperate to get out. Panic was clutching at his throat and stomach.

The loudspeakers crackled. "Right, boys, I've got some good news," came Robbie's voice. "What would you say if I told you that you were being given a little holiday on full pay? The big cheeses have decided to keep you guys on ice there for a while. Just kick back and relax. Anything you want, just let me know. Get some rest. You'll be back down where you belong before you know it."

Dashu coughed, unable to smother the reflex. He pushed his plate away, eyes fixed on the table, then got up and sat on his bunk. Kerr and George both stared at him.

"You all right there, Roland?" Kerr asked.

"Yeah," Dashu said. "Fucking eggs, that's all." But his mind was tumbling. He had to get out. He couldn't carry on. What they were doing wasn't safe. He didn't trust any of this—not these people and not this equipment. He took a pencil and paper and started writing.

ROBBIE READ THE NOTE that came back through the med-lock with the dirty breakfast plates and cursed. Dashu was bailing out. He should have known. This wasn't going to be popular with the management, but there was nothing to be done. Changing him out would be expensive, costing maybe as much as $30,000 in gas and chamber time. The volume of gas in the HRC would be lost, and helium was pricey. Then there was the medical oxygen needed to make up the partial pressure, and the divers' saturation pay on top of that. Keeping the divers inside was a decision Ong had taken and therefore was at his own cost. One of the divers bailing out, on the other hand, was the dive contractor's responsibility. Shit would start to fly. All the same, if someone wanted out there was no point trying to keep him in; it was dangerous to the diver himself and to the others on his team.

Robbie asked for volunteers to escort Dashu out. At least two people had to decompress together so that there would be someone to help if the other had a problem. Jack Ng said he'd do it. Though Ng had been on this same wreck the previous year, this was his first time in saturation. When Dilip heard that Ng was bailing out he was disappointed. Dilip had got him onto the saturation team as a favor—they'd spent time together in the bars of Singapore—and now Ng was throwing the opportunity away. Dashu and Ng were moved into the HRC and sealed off, allowing them to begin their slow depressurization while the other divers remained in the main chamber.

They need not have bothered decompressing early. Within twenty-four hours the plan was changed again. When it became obvious that the storm was not clearing, Spanner decided that this would be a good opportunity to patch up some of the equipment problems that had already started to crop up, and the only way to do so was to decompress the whole system, not just the HRC. All the divers would be coming out after all.

FOR THE NEXT TWO DAYS Dilip rode out the storm aboard *Ena Supply*. The seas were still heavy and he had no chance to try for any of the anchors, but at least Ong could not accuse him of sitting around wasting time.

In Dilip's absence, Winterberg took charge of the barges and the job of fielding the hourly calls from Ong. Though only a barge master on this job, Winterberg was used to running the show. He had been in charge of an offshore commercial outfit in the North Sea for three years and headed up numerous underwater special operations units for oil companies. He patrolled the decks with Germanic efficiency, jotting down long lists of things to attend to on his tiny notepad. He soon alerted Ong to a problem. Ever since *Abex* and *388* had sought shelter in the bay, the water around the barges had turned into a marketplace. Though they were supposed to be alcohol-free, with the vessels moored near the islands the rule had become very hard to enforce. Each day a steady stream of women from the village paddled out to the barges in woven reed coracles, bowl-shaped baskets about the size of a truck tire, which they maneuvered expertly up to the side of the barge. From there they called up to offer their goods—cigarettes, whisky, and "Ba Ba Ba," a local brand of beer.

The artifact desalination tanks on *Abex* were open and unlocked, Winterberg pointed out to Ong. It would be easy for someone to help himself to a piece or two of pottery. The temptation for someone to swap something for a carton of cigarettes or a case of beer might be too great to resist. Ong, silently fuming that Dilip had not brought this to his attention before, ordered a chicken-wire fence topped with barbed wire to be constructed around the storage tanks.

While Winterberg waited for the materials to arrive, he resorted to increasingly desperate measures to keep the girls from coming near the barges. Initially shouting had kept them away, but soon enough they simply giggled at the red-faced, white-haired man and waited until he had gone before approaching again, singing their sales pitch of "Ba Ba Ba!" When Winterberg could stand it no more, he lit the incinerator basket, a cage in which the vessel's rubbish was burned, and ordered the crane driver to lift it over the side. He then swung the boom toward the girls, who scooted off, terrified.

The next day they were back, however, having realized they could outmaneuver the burning basket. Winterberg's eyes narrowed. He went to see Robbie in Dive Control with a proposal. Robbie shrugged. Winterberg went ahead with his plan, waiting until a group of girls sculled downwind of the saturation chambers, still inhabited by the decompressing divers. As the girls passed, Winterberg turned a valve on the side of the system's toilet chamber, unleashing a roar of rapidly decompressing solid waste. The girls flinched at the noise and looked up, blinking and uncomprehending, as a fine, brown mist floated over them. Whether those particular salesgirls returned or not was hard to say, but the coracles continued to be a presence in the waters around the barge even after Winterberg's fence was constructed.

On the evening of May 1, *Ena Supply* rounded the headland at last. On her flat back deck she carried one enormous, mud-plastered anchor. Dilip stood beside it, looking wild and weather-beaten but triumphant. The four crewmen from the tug were covered in grease and mud, looking similarly battle-worn.

Immediately Winterberg and his crew began to ship the anchor back onto *388*. They spooled the wire cable onto the first drum and, using smaller air-powered winches, hauled the anchor off the tug's deck and into the water, then up against the cowcatcher that was mounted on the corner of *388*. As soon as the tug's back deck was clear, Dilip jumped back onto *Ena Supply* and ordered her captain to take them back out to sea.

Dilip and *Ena Supply* battled all night and into the morning, returning with another anchor every seven or eight hours. A ten-foot swell was still running, making it difficult and even dangerous for the crew to hook the pendant buoys that marked the anchors. The men continued fighting the waves and Dilip tore a muscle in his arm during the struggle, but eventually they collected all four anchors. For the second time in a week, Dilip was impressed by the crew's seamanship and congratulated himself on having hired the tug, despite Ong's fury at its cost.

JACK NG AND Roland Dashu came out of the chambers first, the HRC having started its decompression a day earlier than the main system. Ng was relieved to be out. "I ran out of shampoo," he joked, flicking his long black ponytail with a grin. "It's no fun in there. Hard work, no downtime." He had discovered that a twelve-hour shift on the seabed was a world away from the deck decompression diving he'd been used to, where there was always a chance to walk the deck with a cigarette in the evening. He decided to stay on board for the remainder of the job as a deck technician. Dashu disappeared quickly, taking the first supply boat back to the mainland. I never saw him again.

A day later I watched the four remaining divers emerge, stretching and blinking. Though clouds still hid the sun, it was the first daylight they'd seen for a week. They looked casual and unfazed, but Dilip was on high alert. As each of the divers stepped from the chamber he searched them, then immediately went inside. Five minutes later he emerged, cradling something swathed in a T-shirt. I caught his eye and he motioned for me to follow. We walked straight to the barge office.

"Shut the door," said Dilip. "Now, look at this." Slowly and carefully, he unwrapped the bundle on the desk, revealing a beautiful kendi. The high-necked pouring vessel featured lionlike mythological animals stalking toward the breast-shaped spout from both sides, while intricate lotus leaves framed the scene. It was the best example I'd seen yet. "It was in the only place you could stash something in that system. Behind the head, under the floorboards near the bilge," he said.

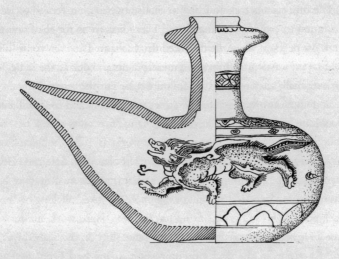

Someone had tipped him off. Dilip took the diver aside and explained once again that no souvenirs were allowed. The licensing for the export permit was still up in the air. Had the economic police, not Dilip, discovered the piece, they could have shut the project down and claimed all of the cargo for Vietnam. The diver nodded. Now that he understood why, he would comply with the rule.

With the system depressurized, the dive crews busied themselves fixing the numerous small problems that had cropped up during the first week. The bent U-bolt mounting that held the bell to its guide wires was replaced. The Environmental Control Unit, responsible for keeping temperature and humidity inside the chambers constant, had been acting up and was overhauled. The helium reclaim circuit, used to recycle the expensive but inert gas that the divers exhaled, had also been giving them problems and was dismantled. Most worrisome, however, were the intermittent problems with the communications. The audio from two of the three Superlite helmets was almost incomprehensible, and one of the cameras was malfunctioning as well. Crews crawled all over the inside and outside of the system, trying to fix everything before the weather cleared and they would be forced to repressurize it, ready or not.

The dive contractors now had to choose two more divers to go into saturation to replace Dashu and Ng. There was no shortage of volunteers. Most, if not all, of the dive technicians wanted one day to be full-time saturation divers, Robbie's son Chad among them. It would be his second saturation, another rung up the ladder to the big league.

More than anyone on board, Robbie knew the dangers involved with the antiquated system, but he did nothing to stop his son. The best he could do was to make sure the equipment worked. "If anything happens to the barge and the escape capsule goes overboard," he said to me after hearing his son's decision, "I'm going with it, right there on top of it."

CHAPTER 17

Ong's Confidence

SIX DAYS AFTER the storm hit, the weather cleared enough for *Tropical 388* to be towed out to the site, leaving *Abex* at anchor in the bay, where it was decided that she would remain for the rest of the excavation. *Ena Supply 3* would be on call twenty-four hours a day, carrying the full crates to *Abex* one by one, waiting alongside while the plastic baskets inside were unloaded, then returning to *388* with the empty crate. Chad joined Kerr and George on Team 1 while another Australian, Troy Harries, was blown down with Team 2. With some experience of the work now under their belts, the divers began to produce fast results. I switched shifts with Mensun so that he could work with dive supervisor Sykes, allowing him to accomplish detailed survey work with greater ease. I began working with Tony Turner. To my surprise, I found that once we'd got used to each other, both the mapping and the recovery work went smoothly.

The next three weeks went without a hitch. Divers were on the seabed twenty-two hours a day and the rate at which the work advanced was astonishing. I was used to site plans taking months, but already a

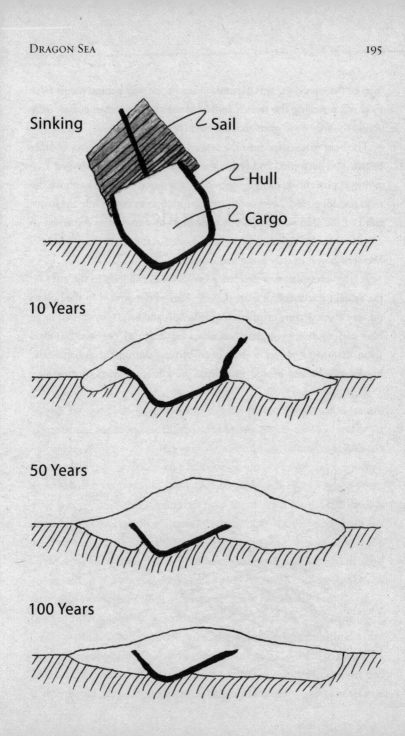

Sinking

Sail

Hull

Cargo

10 Years

50 Years

100 Years

map of the wreck site was expanding across the wall behind me in Dive
Control, revealing the twenty bulkheads of the junk cutting diagonally
across the fifty or so grids that covered the site.

The boat was settled into the seabed, heeled 12 degrees over to starboard. The deep mud had sealed the lower hull, keeping it oxygen-free
and thus preventing its decomposition; any wood protruding above the
mud had long since been eaten away by marine worms. Only the lower
third of the ship was intact, but enough of its extremities remained to
suggest that the bow had been bluff-ended rather than pointed like a
Western bow—exactly as we had expected.

A large proportion of delicate pieces were turning up in the mud on
the vessel's starboard, or eastern, side. Packed uppermost in the hull to
prevent their getting crushed, fragile kendis and ewers, vases, and large
blue-and-white storage jars had cascaded down as the wood around
them disintegrated. Some had been broken during the collapse, the
breaks stained and coated in marine growth from years of exposure.

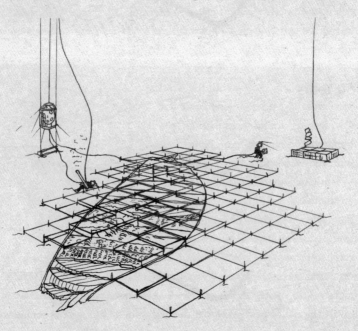

Other pieces showed signs of fresh breaks, evidently caused more recently. Below the mud the stout hull had protected everything it contained, but anything on top or in the spillage area outside had been exposed to the fishermen and their rakes. Luckily, it seemed that few had managed to drag their rakes directly parallel to the hull through the eastern spillage area, for it was still yielding spectacular and delicate pieces that were undamaged.

Within the hull, the cargo had been loaded with the large, heavy dishes stowed lowermost and the lighter rice bowls above, all stacked on their sides, their weight resting on their rims. The gaps between them were filled with cups, jarlets, and animal figurines. We chose to follow the same method when reloading them in their desalination tanks on *Abex,* having added protective nets and padding to prevent chafing.

Working at the starboard end of one of the compartments, a diver uncovered three circles of bowls, one coiled on top of the other in the shape of a cone. I'd seen only straight stacks before, and thinking I was looking at something new, I called in Mensun. When he saw the image on the screen he smiled. This was how bowls were still being transported at the Red River potteries of Bat Trang, the coils bound together with reeds. Though we assumed that the coils within the wreck's cargo must have been bound in a similar way, despite careful examination we found no evidence of them. Being organic and insubstantial, the reeds had evidently decomposed without a trace.

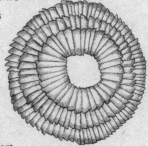

On other vessels, ceramic cargoes were found stowed in straw or even in tea. Apart from occasional packing pieces that had been inserted to prevent stacks from shifting in transit, no evidence of padding was found on the Hoi An wreck, even in areas deep in the hull, where fragile organic materials might have been expected to survive or at least leave a trace. The heavy clay that the divers found around the ceramics toward the bottom of the compartments was the only material that might have served the purpose, serving as padding as well as ballast and commodity.

The method by which the ceramic boxes had been loaded also presented a puzzle. Consisting of a base and a lid, these (mostly circular) storage containers came in a range of diameters, from that of a large coin to that of a small salad bowl. The smallest would have been used for beauty powders or spices among other things, while the largest served as lunch boxes; rice was stored in the base and the meat, fish, or vegetables were placed in a tray that fitted on the rim of the base beneath the lid. The thick ceramic of base and lid would have served as a good insulator, keeping the food warm for hours.

Lid

Decoration on Tray

Reference Mark to
Show Correct Fit
of Lid and Base

Like all the other pieces, the boxes were hand-shaped and hand-painted, and as a result their bases and lids were not interchangeable. We expected that the merchants would have been at pains to keep the pairs together. However, not only were there no traces of binding material (which, as with the coiled bowls, would have decomposed), the paired components of boxes were often found many feet apart even in undisturbed areas within the hull. It seemed that the base and lid had never been packed in a way designed to keep them together, yet reuniting them would have been extremely time-consuming for the traders when they unloaded their cargo for sale. We would discover just how time-consuming once the excavation phase of the project was over.

Both Mensun and I were itching to start a more thorough analysis of the ship's construction. Until then, we'd been frustrated. While the recovery of the ceramics and our map of their distribution continued apace, the structural survey was going painfully slowly. It was difficult even to discern whether or not the vessel had carried a keel. Small details confirmed what we already knew, that she had been built in the South Sea tradition: Planks were joined edge to edge, held together with wooden pins. Marco Polo's testimony that iron nails were used was also proved correct, as was his observation that the seams of junks were caulked with lime. However, even working through Garth Sykes (as opposed to the trickier Tony Turner), Mensun was finding it hard to get accurate information from the divers. First the question went to the supervisor, who adapted it so that the diver could understand—either substituting technical terms or using shorter words that would not be garbled by the poor-quality communications system. The diver then made his own interpretation of the question, carried out the task, and eventually sent his answer back up the chain. We could watch the diver's actions on the monitor, but it was no substitute for actually being on the seabed.

Mapping the structures of shipwrecked hulls is challenging at the best of times. Underwater visibility is often terrible, especially when others are working on the site. Measurements go wrong for no apparent reason and need checking, then rechecking. Currents and surge pull you out of position, tugging at measuring tapes that get snagged when out of sight. The water distorts your vision, making it difficult to judge proportions when sketching. If you're transferring your data onto a map yourself, you develop an intuition for when there's a problem; trying to get someone else to solve it is like describing how to tie a shoelace over the phone. The wide-angle lenses on the video cameras could sometimes pick up details that a diver's eyes would miss. Nonetheless, there were many more cases in which three-dimensional vision was more discerning. Deceptions abound; things that look like features are sometimes incidental while scrapes that look insignificant can offer clues as to how shipwrights turned forests into oceangoing ships.

Underwater mapping had been my favorite part of marine excavations ever since working on the HMS *Agamemnon* in Uruguay. Admiral Horatio Nelson had commanded the *Agamemnon* early in his career, and had always said that she was the ship he remembered most fondly on account of her speed and maneuverability. He had another reason to cherish her decks, for it was upon them that he had met the love of his life, Lady Emma Hamilton. The *Agamemnon* fought at Trafalgar in 1805, the battle where Nelson both died and won his reputation as the world's greatest naval hero. Unlike Nelson, the *Agamemnon* sailed on into old age. Her hull girdled by hemp hawsers to keep her timbers from parting, she had made a last voyage across the Atlantic to join the British Navy's South American squadron in 1809 when disaster struck. Seeking shelter from a storm in Maldonado Bay, the twenty-eight-year-old warship hit a sandbank and began taking on water faster than she could pump it out. Her captain eventually gave orders to abandon ship, and after saving everything they could, the crew took to the boats. According to witnesses every sailor was in tears. For them, losing the *Agamemnon* was like losing Nelson himself all over again.

Exploring her remains in the opaque green water of Maldonado Bay, I felt like an early explorer charting ancient and holy ruins. The sailors had rescued all the valuable cannonry. However, one iron gun remained, buried deep in the mud. The gun was badly concreted from contact with the salt water, but when recovered and treated, the British Navy markings became clear. The numbers stamped into the iron established that it had been on board the *Agamemnon* at Trafalgar. Deep in the bilges that had been worked so hard in the vessel's final minutes, one of the team found a seal used to stamp sealing wax on official documents. It bore both Nelson's name and the starburst motif of the Knights of Bath, awarded to the admiral after his victory at Cape St. Vincent. While the seal was doubtless commemorative rather than having belonged to Nelson himself, it was a poignant discovery, suggesting that one of the weeping old tars had thrown his memento into the sinking hull. For months I measured and plotted the site and never saw more than a few square feet of the wreck, but back on the surface our map grew until we could explore its entirety.

The tropical waters off the remote northern portion of Mozambique allowed a clearer view of the seabed, yet the origin of the ship we were investigating there was obscure. A wooden vessel of early European design had struck a reef near the imposing fortress San Sebastian on Ihla de Moçambique. Carrying a cargo of pepper and porcelain, she had evidently been returning from the East. The style of the porcelain we were recovering told us she had sailed in the late sixteenth century. Other than that, we had few clues as to her identity. Beneath the fifty tons of ballast stones were timbers stouter and more numerous than on any other ship I'd seen. Napoleonic-era warships, such as the *Agamemnon*, used their woodwork as armor and were supported by frames and half-frames that ran up from the keel with gaps between them of only two or three inches. The wreck off San Sebastian was a trader and as such on later vessels her ribs would have been widely spaced. But she had been built with her frames and half frames flush against each other, with no gap between them at all. Such heavy use of timber was not simply an indication of the violent nature of Portuguese exploration. It was often seen on the earliest European ships, when hardwood forests were still plentiful and shipwrights could play safe by overengineering the strength

of the hull. Initially it seemed that, though the vessel had been built following the Portuguese style (such as the manner in which the heads of
the floor timbers and the heels of the futtocks were dovetailed together,
known as knuckling, and the use of lead to caulk the seams), the ship had
not been built in Portugal but in the colonies of Portuguese India. There
were no bronze fastenings, and even iron was used sparingly. Much of
the carpentry was asymmetrical and rough. However, when the wood
samples came back they showed her timbers to be from two oak species
(one of which was cork) that were found in Portugal but not in India.

Floating above the timbers, drawing the intricate details of the ships
that carried so much history, it was easy to forget quite where you
were—until reminded by the wreck's residents. In Uruguay, dives were
often accompanied by the sound of barking in the murk and fleeting
dark shapes. One morning I was head-down in the bilges of the *Agamemnon* when I felt an urgent tugging on my fin. Backing out I turned around,
expecting a fellow diver with a problem, or perhaps an exciting find. Instead I saw nothing but murky green water. Puzzled, I turned back to
the bilge, only to find a sea lion's nose in front of mine. I stared into the
shiny black eyes only inches from my mask. A current was flowing and
I began to keel over, but didn't want to move my hands. I let myself
topple, and the sea lion drifted downward with me, mimicking my fall
and keeping his eyes locked on mine, until at the last minute he snorted
a cloud of bubbles into my face and disappeared. Other members of the
team had found themselves pinned to the seabed by playful eight-foot
males. Evidently the sea lions wanted to see how long these strange
mammals with metallic backs could hold their breath. They, at least, were
playing. At the San Sebastian wreck in Mozambique, I once reached out
to move a ballast stone, but as my hand closed around it spines shot up
between my fingers and a perfectly disguised stonefish darted off; somehow I had not been injected with its lethal poison.

SQUATTING ON *Abex*'s foredeck, Mensun was examining the fluid lines
of the fish pursuing each other across a dish, absorbed by the grace of the

long-dead artist's strokes. He glanced over at the piece to which he was comparing it. It was another fish dish, its creamy-white glaze and rich blue lines jumping out at him against the flaking green paint of the deck.

The methods of depicting waterweeds differed; the dishes were not by the same artist. Mensun felt his mental grip on the Hoi An cargo slip a tiny fraction, his instinct for discerning trends and patterns thwarted. He pushed his shoulders back to stretch, resting his hand on the edge of the black tank beside him while he massaged his back with the other. Hot puddles of sun reflected from the surface of the water, below which he could see the rims of hundreds of other dishes he hadn't yet examined. Next to the open tank beside him was another, then another and another—a line of twenty tanks, each full to the brim with ceramics. Next to that line, another, and then another. Eight ranks of tanks. The responsibility suddenly felt crushing. How could he hope to analyze and present such an enormous haul? What had so drawn Mensun to this cargo in the first place now intimidated him. He could feel the watchful eyes of the MARE board members that had befriended or doubted him. He did not want to let down the former or justify the latter.

Beyond the crowded decks, through the gaps in the islands, was the open South China Sea. Out there, as in the bay, the water was oily-flat. It wouldn't stay that way for long: They'd soon be entering typhoon season. The previous two years it had come early. Mensun knew it was vital to keep information coming in from all fronts, both from seabed surveys and from the draftsmen and the photographer. The whole project could still collapse and if it did, he would be left without support. That was the problem with trying to keep the recovery phase separate from

the analysis. Were something to go wrong, he could find himself without a team or any other resources with which to work on the artifacts that had been recovered. That was his constant fear—and the reason why he refused to leave all analysis until the operational phase was over, as Ong had wanted.

When the ship had sailed—and thus when her precious cargo had been manufactured—was rapidly becoming the most pressing question he had to answer. Before the first season had even started, in 1998, Mensun had visited London's Victoria and Albert Museum to meet John Guy, curator of Southeast Asian Ceramics. Guy had co-authored one of the only textbooks dedicated to Vietnamese wares, *Vietnamese Ceramics: A Separate Tradition,* and Mensun and Ong had decided to seek his opinion as to the date of some pieces. Though Mensun had yet to begin his crash course in Southeast Asian ceramic studies, reading the book he had sensed an uncertainty about the dating. When they met, Mensun had asked for clarification but found Guy hesitant, as if conceding there was a lack of solid evidence. Two years later, and with data now flooding in, one of Mensun's main hopes for the Hoi An excavation was to clarify and reveal exactly when this style of ceramic had been created.

There were some standard tools he could use. Dendrochronology of the ship's timbers might eventually identify the year in which the trees from which they were made had been felled. Trees grow more in wet years than they do in dry years, and the changes are reflected in the rings in the cross-section of a tree trunk. Over time, the pattern of dry and wet years becomes as distinctive as a bar code, and can be used to identify when the tree was alive. As long as the origin of the tree is known, its growth rings can be compared with those from trees of known age from the same area, and so the sample's age determined. Most regions of Europe and America had enough reference material and climatic data to permit great accuracy, but by comparison the Southeast Asian database was sparse. Getting a match would be a long shot. If the tree rings did yield a date, Mensun would have to add perhaps five years for the wood to season, and another one or two for the ship to be built. Then he would have to estimate how long the ship had been sailing before she'd sunk. So far he had seen nothing to suggest this was an old ship, though he had noted some signs of repair on the outer planking.

Finding Chinese porcelain as part of a wreck's cargo is always a strong indication of its vintage, both because the stylistic progression has been so thoroughly documented and because certain pieces were stamped on the underside with the mark of the reign in which they were produced. By contrast, the rarity of Vietnamese ceramics and the absence of reign marks meant they could only be dated based on similarities to Chinese designs and technologies, and the diffusion rate of cross-border influence was unknown. Indeed, Mensun had a theory that the Ming Empire—which had perpetrated the brutal occupation at the beginning of the fifteenth century—was so detested by the Vietnamese that its style was ignored in favor of that of their predecessors, the Yuan. The Yuan were regarded in a more friendly light because they were not the long-detested Han Chinese but were Mongols descended from Genghis Khan. They had invaded China in 1271, but peasant revolts eventually ousted them in 1368 and led to the start of the mighty Ming Empire.

The idea of the Vietnamese harking back to the Yuan for inspiration appealed to Mensun. He knew how strongly resentment could persist in the heart of a nation that had been violated. When Argentina had invaded his home, the small British colony of the Falkland Islands, in 1982, patriotic ardor had flared within him. Fifteen years later the thought of Argentina still made him bristle, even when we were working in neighboring Uruguay. On some level Mensun identified with the rebellious sentiment behind the reluctance of the Vietnamese to follow the Chinese tradition, to support the influence of the Mongols that had invaded China. The more he thought about it, the more sense it made that the cargo had been crafted far earlier than John Guy claimed: not in the early sixteenth century but in the mid-fifteenth century, in the heart of Vietnam's brief golden age.

A few other wrecks had been found in the South China Sea whose cargoes of Chinese porcelain had been mixed in with Vietnamese wares, providing a valuable key. On the Pandanan wreck, for example, discovered in 1995 off the Philippines, a few pieces of Chu Dau ceramic (the same source as those from the Hoi An wreck) had been found among pottery from China, Thailand, and central Vietnam. The Chinese wares dated from the mid-fifteenth century, while the central Vietnamese ceramics were from Binh Dinh, near the Cham capital of Vijaya, which was sacked by the northern Vietnamese in 1471, suggesting that the ship had sailed before then.

The ceramics on the Hoi An wreck were not exclusively Vietnamese. Scattered around in the area thought to have been the galley were some Chinese bowls. Their thin walls and regimented designs stood out from among the Vietnamese wares. They were also distinguished by their chips and cracks. These were bowls that had seen a lot of use, and were evidently personal possessions rather than cargo. All three were from the Jingdezhen kilns and appeared to date from the interregnum period between the early and late Ming emperors (1435–1465). The bowls might have been in use for some time, blurring their accuracy as a dating tool, but they were a start. They also suggested that some of the crew who'd sailed this ill-starred junk were Chinese. On the other hand, large storage jars recovered from the starboard spill area implied otherwise, for they were from the Ban Rachan kilns of Singburi province in Siam (present-day Thailand). Some were found containing the bones of red snapper and albacore, though not in sufficient quantities to have been cargo. Others had been lashed to the deck (judging from their location on the wreck site), to provide the crew with fresh water. Perhaps the sailors were Thai merchants who had visited China? Or Chinese merchants with a Thai crew?

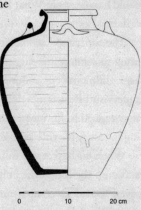

It worried Mensun that so few other items from the ship or her crew had been found so far. The galley area had produced some copper cauldrons with charred bases, and a set of nest weights. Scattered finds had turned up from elsewhere, including two mirror handles and an inkstick grinding tablet (used by calligraphers to prepare their inks), some knives, buckets, a gong and other metal wares, but nowhere near as much as he would have expected. The upper sections of the hull (where the living quarters were) would have been vulnerable to dispersal by the rakes, but Mensun couldn't shake the fear that despite his instructions, the divers were ignoring anything that wasn't ceramic. Either they were simply not spotting the less obvious shapes of the nonceramic items or were deliberately leaving them, the result of conflicting orders coming from his codirector. He was beginning to suspect the latter.

KERR WASN'T focusing on anything except loading dishes. He was deep inside one of the compartments in the center of the wreck, and not happy. It was hard work; he was only halfway through his shift and already the back of his neck, his shoulders, and his lower back were aching from the repetitive cycle of movements involved in prying dishes from the thick clay, packing a basket, and transporting it to the recovery crate. But that wasn't what was making him angry; the pain and the labor he could handle easily. What Kerr didn't like was being forced to follow bad orders.

The supervisors wouldn't listen to him, and so he had stopped bothering to send up his suggestions. His previous role as operations manager didn't mean a thing now that he was in saturation; to the supervisors he was merely a grunt. He could hear it in their voices when they replied to his ideas; they thought he was too big for his boots, trying to justify his title as chief diver.

When his basket was full of dishes, Kerr heaved it up above his head and onto the top of the bulkhead. Then he pulled himself up from the seven-foot-deep compartment and through the grid, out of the opaque grayish-brown water and into the current. He had sent up countless proposals for how to speed up the work. From the very beginning, for example, he had advised against laying the grids at an angle to the bulkheads. Why not line them up and keep everything simple? But because no one had listened to him, the entire site was now crisscrossed by grids running at an angle to the tops of the compartments. He put his hand around to feel for his umbilical, and then checked that he had come out of the same grid he had dropped in through. Come out of the wrong one and he'd be snagged. If anything went wrong, his quickest way back to the bell would then involve unthreading himself through the maddening labyrinth of crisscrossing iron bars.

Nor did he understand why the supervisors insisted on dropping the recovery crate on one side of the wreck and then sending him to work on the other. Kerr was spending his whole time picking his way across the grids with baskets of ceramics. The excavated compartments formed chasms beneath them, so when he traversed the site he had to try and place his feet on the grids themselves. They were made of sharp angle-iron, and though originally they had carried high-visibility paintwork, it had all peeled off within the first couple of weeks. With two divers using airlifts and water jets the water was murky, and with the grids now all the same rust-brown they were impossible to see.

Wherever he went, his umbilical cable trailed behind him. Slightly negatively buoyant, it settled amongst the scaffolding legs of the grids and got caught, yanking him off his feet when it did. Trying to make his way across the grids in the current was like trying to clamber through a climbing frame on a moonless night with a gale blowing, wearing full dive gear, trailing a cable, and carrying a heavy basket of fragile ceramics. Every slip or misstep was punished with a blow or a scrape to the shin from the angle-iron, tearing his waterlogged skin like wet paper.

The lacerated shins joined the welts on his neck, caused by the helmet when he bent over to do grid scans for the archeologists, and the burns. The hot-water unit on the surface was malfunctioning, occasionally sending down bursts of scalding liquid that came searing through the heating tubes of his suit, scorching him before he had a chance to shut off the valve. Neither the burns nor the cuts would heal. It was like being in the jungle, where nothing closed up and everything got infected. Now his mind was beginning to feel the same, with resentment and annoyance seething. Despite the fact that he had found this wreck, and that he had run the first survey of it, he was being treated like a stupid diver.

THE DOOR at my end of Dive Control cracked open. Ong's face appeared. It was a shock to see him. Though we'd spoken on the satellite phone, he had been off the barge for almost three weeks.

"Ah! Good morning. How's it going?"

"Fine," I replied. "Nearly finished another compartment, starting to understand that stern section a bit more."

"Good. Can I come in?"

I smiled, embarrassed. It was his barge; he owned it. Ong sat down, looking around the archeological annex of the Dive Control room, nodding with satisfaction. The whiteboard showed a schematic of the grids and their numbers together with the labeled bulkheads of the wreck. A code of colored dots marked which grids had been completed and which were in progress, while the monitors in front of me showed both divers at work on the seabed.

"So, at last we're getting somewhere, *mmm*? Feels like we've finally got this job under control." He turned and stared me straight in the eye, still nodding gently. I nodded back, agreeing. In a matter of seconds, Ong had made me feel as if the entire success of the project was up to me.

"Look, I've been thinking," he said after a pause, leaning back in his chair. "This wreck is bigger, much bigger than we'd thought. We've already got more than one hundred thousand pots on *Abex,* and we've not even touched half of the compartments, right?"

"That's right. We've reached the bottom of only two of the compartments, and there are eighteen. There's definitely another season's worth in there, or two. We can go away and look at what we've got, then come back and have another go next year."

Ong shifted in his chair, leaning forward as if I'd just suggested something to him. "Why wait?" he asked. "What I am thinking is that if things are going well, why stop now?"

"Well, typhoon season kicks off in June, so that's an enforced departure date."

"That's when typhoons can come, sure. But as we've discovered, they don't necessarily stick to the timetable. If we get a break—and at the moment it seems like we have—we should make use of it, right? Really the only thing stopping us staying is the fact that we're running out of space on *Abex.* We can't physically store any more ceramics!"

I nodded slowly, not yet following his line of thought.

"But that's easily solved," Ong continued. "We just bring in another barge."

I blinked. Another barge? Ong nodded, his eyes shining. He drew his laptop out of his bag as smoothly as if he were taking his hand from his pocket and opened a spreadsheet. "From the figures Dilip's been sending me I can see we're recovering between two and five thousand pieces per shift," he began. "This has been okay, so far, but how much of the cargo do you think remains?"

That was a difficult question to answer. I reached for a recent print-out of the site plan. "Very roughly, we can say that seventy-five percent of what we've got so far has come from the compartments inside the hull, with the majority of the remainder coming from off the starboard side. That seventy-five percent, say eighty thousand pieces, has come from less than two compartments. So that's forty thousand per compartment. There are eighteen compartments, and all appear to contain ceramics, so that's about, what, seven hundred and twenty thousand pieces. And that's not counting everything that's outside the hull, which could be extensive."

"*Mmm.*" Ong was nodding his head. "So you're telling me you think there might be, what, three-quarters of a million, maybe a million pieces on board?"

"It seems incredible, but I think so. Mensun might have a better . . ."

"I'm asking you." Ong's expression had sharpened at the mention of Mensun's name.

"Okay . . . ," I replied, slightly taken aback. "I'd say around that then, very roughly."

"As you know, I badly need the archeological aspect to make this project work, but I also badly need to have a substantial cargo to sell at auction. From what I've seen things seem to be bottlenecking on *Abex*." He went on to explain that the crates were arriving faster than the crews there were able to unpack, label, and move the ceramics into desalination. There was a lot of unnecessary recording being done, he felt, diverting attention from the operational work. We needed to get the recovery done now, the recording later—just as we'd planned. "If we don't make money or at least cover our costs, we can forget about coming back here or taking a saturation barge anywhere else. I think Mensun has trouble seeing this."

"There's a certain amount we have to do at the time, Mr. Ong, or the information will be lost," I said, feeling defensive.

"Like what?" Ong replied, his eyes locked onto mine. "Look. If we do this the way we planned we lose nothing. We—you and me—spent a long time thinking about a system that preserves the information of exactly where everything comes from but which postpones the time-consuming recording phase until later! Am I wrong?"

He didn't wait for an answer. "Listen. I know this isn't how you're used to operating, but look around you. Look at all this kit, all these people. This is not a summer school excavation in some field. These aren't volunteers. Somehow I've got to pay for all this. I want this to work for both sides, but I'm not a charity. This thing has got to be commercially viable! We've had a rough ride so far. If we're going to bring in another barge we need to know that our systems are working."

He had a point. With meticulous detail, Ong had unpicked the processes involved in tracing where artifacts come from, from first uncovering to final storage. I was the one that had helped him plot it all out in Kuala Lumpur while we'd prepared for the job, not Mensun. For us to recover artifacts at the rate that we had agreed upon, we needed to use such systems, but Mensun didn't seem to appreciate this. He wanted to work the way he'd always worked, but on this job that wouldn't be possible. If this was the future of MARE, we'd have to find a way to work efficiently. I decided I had to go over to *Abex* to see Mensun (where he had been working for the last few days) and confront him.

TWO DAYS LATER, Robbie was sitting in his captain's chair, pondering the news that Dilip had just delivered: The job was being extended by at least another month. His eyes stayed on the gas control panel, unconsciously checking each of the gauges for the hundredth time that day. He knew only too well the simultaneously lethal yet life-sustaining pressures that coursed around the pipe-work. His gaze wandered involuntarily to one particular monitor whose flickering black-and-white screen showed his son's sleeping form. What would Chad make of the news? Normally, extending the job past one month would mean swapping out the divers and bringing in new ones, but doing so would cost time and money. Rather than decompressing the current divers, Dilip wanted to know if they would extend their saturation. Looking at his son's pallid face, Robbie could see he was already exhausted.

In other parts of the world, keeping the divers inside any longer would be illegal. In the Norwegian oil fields, for example, divers are allowed to stay inside for only fourteen days. The British Health and Safety Executive's limit is twenty-eight days, and this is used as the standard in the North Sea and throughout most of the world. But this was the South China Sea, and they weren't in a regulated oil field. The rules did not apply.

Living at extreme pressures exerts a profound influence on human physiology, in ways that are still little understood. Not all of them are bad. Indeed, hyperbaric medicine is a rapidly growing new industry. At pressure, a liquid can absorb more gas than normal—as demonstrated when soft drinks are bottled under pressure, absorbing gas that later fizzes out when the lid is popped. In a similar way, blood plasma, synovial fluid, and cerebral fluid are all able to absorb more oxygen in a high-pressure environment. As long as the pressure and duration don't push the oxygen to toxic levels, the results can be beneficial. With more oxygen in the body's fluids, more of it reaches traumatized or infected tissues—especially important since damaged areas already suffer from decreased blood circulation and therefore receive less oxygen. The toxic properties of oxygen also play a role, as many viruses and bacteria cannot tolerate it at the increased concentration that is received under pressure. As a result, patients placed in oxygen-enriched pressure chambers repeatedly for two or three hours at a stretch find wounds and infections heal faster, some cancer treatments can become more effective, and even wombs become more receptive to implanted eggs.

Twenty-eight days is a different matter. Deprived of natural light and constantly damp from the water and the humid chamber, the skin yellows and bruises easily. As tiredness sets in, the immune system begins to fail. The sealed environment is hard to keep clean, transmitting infections easily. The hot water trickling inside the divers' suits against

their skin for ten hours a day is drawn from the area immediately surrounding the barge, unfiltered. Dilip had once been underwater when
the barge he was working from passed through a smack of jellyfish.
Their stinging cells were chopped up as they were sucked through the
impeller of the heating unit, then pumped down the umbilical straight
into his suit with the hot water, covering his body in welts within seconds. Whenever the hot-water system intake ended up down-current
or downwind of the ablution outlets, the barge's human waste traveled
the same route. Not that any external factor was needed for infection in
the South China Sea. The seabed itself contained a high concentration
of naturally occurring arsenic that was slowly poisoning the divers on its
own as it leaked through their suits and broken skin over their long
hours on the bottom.

As if the long hours combined with tasteless food didn't take enough
of a toll on the divers' bodies, they also had to contend with a "negative
nitrogen balance" created by high pressure. Within the human body nitrogen occurs naturally only in protein, approximately 70 percent of
which is contained in the muscles. High pressure causes the body to ex-
crete more nitrogen than it can absorb (a negative balance), and as a result it begins to absorb its own proteins, starting with the muscles before
moving on to the vital organs, where the remainder of the body's nitrogen is stored. In Team 2, Mike Hughes—a diver from New Zealand
with twenty years of saturation experience (amounting to more than
five years lived continually at pressure)—took steps to combat this by
eating two raw eggs with every meal to keep his protein levels high.
Though it earned him the nickname "Fat Boy," he gained an advantage
beyond combating the negative nitrogen balance: He could spend the
entire shift on the seabed without using the temperamental hot-water
system, relying only on exertion and his own insulation to keep himself
warm.

The long hours of operation were also affecting the equipment. The
technicians were being forced to play catch-up with the repairs, and that
wasn't a good state for a life-support system to be in. The problems with

the Environmental Control Unit were getting worse. The equipment was having trouble cooling the chambers down in the semitropical sun and was breaking down frequently. High temperatures inside the system were a big concern for Robbie. The highly conductive helium atmosphere inside meant the divers would overheat rapidly, while the crushing pressure disabled their bodies' natural protection systems. Heatstroke had become a real concern.

Then there was the communications system. The technicians' previous repair efforts hadn't worked. The two feeds from the helmets were interfering, no matter what they did. The only way around the problem was to keep a single channel open, which meant talking to only one diver at a time. The other diver was left in silence. This usually meant that he was working at some mundane task not requiring instruction or feedback. Were something to go wrong, his only recourse was to wave his hand in front of his helmet camera and attract the supervisor's attention, who could then switch his input to that diver. Having one dud communications channel wasn't ideal, but for the time being they would have to make do.

After Robbie made the announcement about Ong's decision to extend the job, each of the men inside the system was given a day to think about whether he wanted to extend his saturation until the wreck was finished. The incentive dangled in front of them was enticing: continued pay of between $500 and $800 per day, with a bonus on top. The question was how much more their bodies and minds could take. Robbie wondered what his son would decide. Chad had been on Kerr and George's team, and he could see the strain it was putting him under. Robbie half wished his son would call it a day and come out onto the safety of the deck, but part of him also wanted Chad to show his mettle and prove he was his father's son.

The divers unanimously decided to stay inside.

CHAPTER 18

Dividing Loyalties

THE NEXT CHANCE that I got, I joined *Ena Supply* on one of its daily taxi journeys, carrying a crate full of ceramics back to *Abex* in Cu Lao Cham Bay. I was on a different mission: Mensun had to be made to realize that we needed to be working in accordance with the plan. As we drew up alongside *Abex*, Vietnamese artifact-cleaning crews and Indonesian deck hands leaped as one onto the back deck of the tug. Some began untying the lids of the recovery crate while others put down ramps. The numbered baskets were unpacked from their compartments and handed down a chain of men, each man's eyes flicking over the contents as they passed. Nothing could be taken from its basket for fear of losing track of where it had come from, but shouts went up when an especially beautiful or intricate piece became visible. Large storage jars were transported independently, two men bearing the weight of the muddy contents, another gift waiting to be unwrapped. *Abex*'s deck was six feet higher than the stern of *Ena Supply*, and one by one baskets were slid up a ramp slick with mud and seawater, picked up on *Abex* and carried to the sluicing tables, where they were stacked in different areas.

Once empty, the recovery crate was restocked with empty baskets and *Ena Supply* could cast off and return to *388*, to await her next consignment. The washing crews then took up their stations at the tables, taking basket after basket from the stacks and carefully washing off the mud. Each table was watched over by a supervisor from the Vietnamese National History museum or the Institute of Archaeology; they were concerned about overenthusiastic cleaning that might damage delicate pieces and were keeping an eye out for designs that had not yet been recorded. Mensun had alerted them to be on the lookout for depictions of people—which were rare—and to check underneath each dish. Occasionally, beneath the swirl of brown glaze that usually covered the base of each piece was an inscription in Chinese characters that carried a Vietnamese meaning (the Vietnamese were still using Chinese ideographs to write their language). Aside from its intrinsic interest, Mensun hoped that an inscription might suggest the date when the piece had been painted. So far no dates had turned up, but the base of one dish did bear a line that could be interpreted as: "Whatever winds may blow they will not deter the artist in his pursuit of beauty."

Other pieces featured trial sketches on their undersides suggesting the painter was experimenting with a new idea, while some testified to the use of child labor in the workshops; tiny handprints were occasionally found imprinted in the brown glaze, evidence that children had carried the piece when it was still wet. When a table supervisor noticed one of these items he would put it aside and give it its own number, leaving the rest to be stored in separate desalination tanks according to their find location. If a new design was found a cry went up, and supervisors and washers crowded around it. The bank of known designs was constantly increasing, and a wall of honor charted who was spotting the most.

Occasional breaks were called, and the chaos was suspended while mud-streaked, sweat-soaked washers sat on empty upturned baskets to drink their tea. They couldn't sit for long. In order to keep up with the divers, they had to process all the baskets before the next crate arrived, which could be within eight hours of the previous crate. Magnus Dennis was constantly in the middle of the action, directing the madness, shouting and laughing. Watching the washers, I felt a stab of envy. Suddenly this felt like the heart of the whole project. They were actually getting to examine the cargo, to caress and smell it. The atmosphere could not have been more different from the factionalized and equipment-intensive shift work on *388*.

Among the streams of workers was Ivo. He greeted me enthusiastically. "Ah! So you come to visit us at last! Wonderful! We have much to show you, lots of exciting things. Look!" He began scrabbling around in a plastic pot. From inside he finally pulled two faceted pieces of stone. "Gaming pieces! They were found in the mud which we are sifting! We don't know what the game was, exactly, but—"

I was confused. Sifting? That wasn't part of the plan. It also happened to be one of the worst jobs on an excavation. I myself had spent many tedious hours bent over a sieve, sifting through mud and grit, looking for anything that the divers might have missed on the seabed while working with airlifts. Filter bags had been fitted over the ends of the airlifts, but the plan was to go through them *after* the excavation.

"The mud you're sifting?" I asked.

"Yes! But please—do not tell Mr. Ong. I am supposed to only be a translator." Ivo looked genuinely worried. "He doesn't like me to do extra work like this. Come, I will show you." At the end of one of the long washing tables was a large mound of mud. Behind it were tanks containing other mounds. Two Vietnamese—an archeologist from the National Museum and a member of the Border Guard—were slowly sifting through it, using trickles of water from hose pipes.

Ivo had been picking through the mud on his own in the evenings and gradually had accumulated a dedicated following of helpers. In part this was because he was admired by everyone on the barge. He spoke all their languages (Malay, Indonesian, Vietnamese, as well as at least nine others), sometimes even better than they did themselves. He also knew the history of their cultures. Ivo was especially popular with the Vietnamese, whose language he begun studying while a graduate student in Prague. Indeed, some of the senior Vietnamese archeologists admitted that the project had met less bureaucratic resistance because they'd felt reassured by Ivo's presence, assuming that he wouldn't be involved with unethical people. However, it wasn't simply Ivo's encyclopedic knowledge and gentle ways that won their hearts; it was because he was Czech. The Czechs were among the first Europeans (aside from Marco Polo, of course) to go to Vietnam. Orderick of Pordinone, an Italian-trained missionary from a Czech family, visited Champa at the end of the thirteenth century. There he witnessed the king's army and its thousand elephants. A link between the two nations persisted. "If you cut a Czech's finger," it was once said, "you will find he bleeds like the Vietnamese."

As the Vietnam War was escalating in August 1966, the prime minister of then-communist Czechoslovakia took a delegation to Hanoi (which was already being bombed sporadically) to determine how aid could best be delivered, and to discuss North Vietnam's political strategy. Though a scholar rather than an official government interpreter, Ivo was asked to join the delegation to translate key public speeches, including the Czech prime minister's official address.

Ivo told me that at the reception before the main event he had stood behind the Czech delegation's table, clutching his translation and keeping a low profile. He was watching Ho Chi Minh move between the various delegates, among them high commissioners and other diplomats from Britain, France, China, and the Soviet Union. Ivo followed the great man's progress with interest, not least because once the hand-shaking was finished Ivo would have to stand up with the prime minister and make his speech. Ho Chi Minh was halfway across the hall when he turned and began making his way back toward Ivo, picking up two glasses of Champagne on the way. He walked straight up to the Czech and presented him with one of the glasses. "*A votre santé, camarade,*" Ho Chi Minh said to Ivo, raising his glass.

"*A votre santé, monsieur le président,*" replied the astonished Ivo. Despite being the leader of a country at war and having a room full of international dignitaries to meet, the president had spotted the abashed-looking translator without a drink and crossed the reception hall to greet him.

In fact, Ivo's role during that trip had been important. To save time, the Czech prime minister had decided to speak only the first and last sentences of his address, leaving Ivo to deliver the main body of the speech in Vietnamese. Ivo had no dictionary with him, and had been given the final draft of the speech only the previous evening. He admitted to me that "there were some stupid little influences of Czech in my Vietnamese . . . I called the 'Central Executive Committee of the Communist Party' just the 'Central Committee,'" Ivo recalled the event, still cringing. "But you've got to say it in full."

By now I had grown to appreciate the depth of the Vietnamese affection for long committee names. Ivo's slip-up apparently hadn't offended Ho Chi Minh, who was so impressed by the speech that afterward he invited Ivo to join him as his personal guest at the official dinner, seating the young scholar on his left. Ivo said little but listened to the Vietnamese president intently all evening and learned, among other things, the difficult feat of eating kaki fruit without getting covered in its flesh. They got on so well that every day for the rest of the diplomatic visit, Ivo was summoned to spend time with "Uncle Ho," despite the Czech prime minister's complaints that he was occupying too much of the Vietnamese leader's time. Ivo went on to publish a Czech translation of

Ho's *Prison Diaries,* written during his imprisonment in the south of China between 1942 and 1943.

It was refreshing and inspiring to watch the *Abex* crews at work. On *388* all the talk was of numbers—of intact pieces, dive times, and over-time. It felt like a business; people relaxed when they knocked off shift. On *Abex* they used their time off to do more work, and Ivo, by far their oldest hand, was putting himself out more than anyone. It reminded me of the early MARE excavations, which were staffed solely by volunteers. Somehow, by paying for our own flights and buying our own food, it seemed a waste to spend time doing anything other than the archeology which we had come to do. With two or three dives each day (depending on the depth of the site), every waking hour was spent recording our ac-tivities and discoveries, bonding the teams into tight-knit groups.

Once the unpacking of the crate finally seemed under control, I pulled Magnus aside and asked if he knew where Mensun was. "Good you asked," he said as we walked away from the crews. "I've been want-ing to talk to you. Not something I could really say on the radio, if you get my drift. Now I don't mind, I'll do whatever you guys want, but I just want to check what the score is. It's getting frustrating having to re-peat stuff all the time."

I didn't understand. With *Tropical 388* so far away, my understanding of what was happening on *Abex* was remote.

"Like taking pictures. Mensun keeps asking me to shoot rolls for him. Now as far as I'm concerned they're all for him. Well, for him and Ong together. But Mensun wants these ones on the sly. Says that he'll never see the pictures again if Ong gets the rolls. Now I love Mensun dearly, but Ong's the one that's paying for the films and for everything else. So Mensun gives me the film and says 'There you go' but it's not the point. I could double up these shots of the pottery so easy by just whacking out two frames of every angle, then they could just divvy up the slides. But taking separate rolls means setting everything up twice, so it takes twice as long. Again, I wouldn't mind but I'm falling behind."

I nodded, feeling my heart sink.

"And that's just the photos," he continued. "It's going on with other stuff too. What do you think I should do?"

I FOUND MENSUN sitting among the desalination tanks at the far end of *Abex*, deep in discussion with Dr. Tong Trung Tin, a senior Vietnamese academic and specialist in tenth- to fifteenth-century ceramics and architecture. Nguyen Viet Cuong, a young Vietnamese archeologist, was breaking open artifact bags for them. The washing crews had labeled the net bags according to when and where they'd been discovered—in which grid and layer. Our agreement with Ong, of course, had been to leave them all in desalination until after the excavation. The deck around Mensun's chair was covered with little groups of jars. We talked for a few minutes before I could speak to Mensun alone. I gestured at the ceramics and asked him what he was working on.

"Oh, putting together groups for the draftsmen," Mensun replied. "They're also good study collections for us to start getting a handle on this lot."

I looked down at the pots.

"Isn't this something we could do later on? We planned things a certain way, and it messes up our systems if you're taking pieces out of storage and mixing them around."

"They're tagged, Frank," Mensun replied, the tiniest irritation flashing behind his eyes.

"I know, I'm sure. It's just the time—"

"Look, trust me, I know what I'm doing," Mensun said, then fell silent as he leaned forward in his chair and began to get up, wincing. When he got to his feet he was still bent over. Due to the combination of hard bunks and lifting stacks of heavy ceramics all day, Mensun had slipped a disc in his back. Determined to ignore it, he lived on a diet of morphine-based painkillers. Every day he was more stooped and careful in his movements. It was a stark contrast to Ong, who grew ever more fit, channeling his anxieties into a rigorous regime of pumping weights and a strict diet of chicken and tomatoes.

Holding a hand to the small of his back, Mensun slowly stretched upright. "Let's go and grab a coffee." He passed Cuong a list of requests, telling him to find what he could from it, and then asked if I wouldn't mind carrying a large stack of dishes, each bearing a numbered tag in its well.

"You look tired, Frank," Mensun said to me as we made our way between the tanks. "Everything going okay over there?"

"Fine, fine."

"Ong getting on top of you?"

"No. Actually we've been having interesting discussions."

"Really? He's been prowling 'round here, upsetting everyone. I just don't understand him. Just when things are going smoothly. He just can't let it alone."

"He only wants to do things systematically," I countered, "and I sympathize with that." If we were not sacrificing any information by recording the artifacts afterward, why couldn't we bend on that issue and let the *Abex* crews concentrate on packing the pieces from crate to desalination tank quickly so as to keep the divers working efficiently?

We were passing an area covered by an awning. Mensun pushed inside through the canvas flaps. Inside Magnus was back at work, standing next to two cameras on tripods facing a fully lit photo stage. He was surrounded by piles of artifacts. I noticed for the first time that he had dark circles beneath his eyes from lack of sleep. Still, when I handed over the stack of dishes I was carrying for Mensun, he saluted with a knowing smile.

"Listen, you've got to trust me," Mensun continued as we carried on walking aft. "I've been doing these excavations for a long time. I know this one is different in many ways, but there are some things that remain the same. We're here to collect all the information we possibly can from this wreck. As you know, that involves trying to do lots of different things at once. The ceramics, they're obviously our main priority. But there's also the ship itself, how she's made. It may be a good thing for the washing crews on *Abex* to take their time—it gives us an excuse to do more survey work on the seabed. We need to think carefully about how much data we can realistically get on the hull, what information is key, and what we can get that will be accurate.

"The ship's fittings," he continued, "and the crew's personal possessions. We're not finding enough of them, and that makes me worry. Have we drummed it into the divers enough how important those artifacts could be? Ivo's doing vital work on the washing tables, but we need to keep a good lookout on the seabed, especially now we're down deep in the holds. That's a classic place for coins to accumulate. I'm relying very much on your experience here, to help us get what we need from this."

"I'm trying," I replied. "My problem is that I feel like we're repeating a lot of work for ourselves up here on *Abex*, that we're not doing things efficiently. Like taking stuff out of storage to study it out here, taking photos . . ."

"He has got to you, hasn't he? That's the fourth time you've said 'efficiently' in so many minutes," Mensun said with a grin.

"No, I just think we should stick to the plan. We're all in the same boat, aren't we? We all want this to work—why can't we pull in the same direction?"

"Look. We've got to keep our priorities straight. We're here to do archeology, and archeology is not about efficiency. You've heard from Ong how precarious the funding situation is." All it would take was for one of the brothers to get scared and withdraw his support, and the whole operation would come crashing down, Mensun explained. One day we'd be going at full pelt, the next we'd be left with nothing.

"We can't do the recovery first and then the archeology," he said. It didn't work like that, for one fed the other. And if the project collapsed after we'd finished the recovery phase, then Ong would have his pots but we'd be left with nothing. We'd have deserted our duty. We had to grab everything we can from these pots now, build our knowledge incrementally. It might not be the most efficient way but it was the most practical, given Mensun's experiences. "If Ong doesn't like it, that's tough. I'm the excavation director, and without my say-so Ong would not be here. I could end this all now. All it would take is one word from me that I wasn't happy about the archeological standards.

"Now, I don't want that any more than you do, but we have to be free to do our job. This is so important, Frank. You've been stuck away on that barge over there, surrounded by the commercial divers. They don't understand this stuff—all they understand is money. But look around here. Listen to the Vietnamese. None of their archeologists can believe what they're seeing. They can tell this is the most important discovery in their history. It's our job to make sure that this remains about them, not about the money, the profit, or the efficiency."

His arguments were valid, but I felt disoriented, unable to discern what was right and what was wrong. I understood both men's perspectives and that the trust between them had evaporated. Mensun didn't have confidence that Ong would follow through with the post-excavation, and so was trying to get all he could as we went along. Ong, on the other hand, thought Mensun would try to subvert his schedules with extra archeological work. What seed their paranoia had grown from I didn't know, but it was now feeding upon itself. Trust was integral to cooperation, but cooperation, I had started to appreciate, meant compromise. With the Hoi An wreck beginning to show itself to be more significant in terms of archeology and art history than Mensun had dreamed, and of greater financial potential than Ong had dared hope, neither man wanted to meet the other halfway.

"MENSUN! FRANK!" Ivo came bustling toward us from the bow, holding up something in his fingers. "Look what I have found! Eggshell, and almost intact!" He was carrying a small, white fragment from the base of a cup, the upper portion of which had broken away. He placed it in Mensun's hand.

"Wow, Ivo. So you were right . . . Amazing! Look," Mensun said, handing it over to me. It seemed impossible that such a delicate piece of eggshell-thin porcelain could have survived both the sinking and the raking.

Ivo had been the first to notice the small, thin white fragments on the original survey, and insisted that they be registered as artifacts (to the bemusement of the Vietnamese archeologists, who assured him that the fragments were either glaze that had come away from another piece or were an animal's shell). The next year Ivo had discovered bigger shards of the same type. Though they were mixed in with the rest of the cargo, the archeologists assumed that they had to be Chinese. As far as they knew, the ancient Vietnamese had neither the materials nor the expertise to make such fine-walled wares.

"Now, the really special thing." Ivo extended his hand out to take back the tiny cup and held it up toward the sun. His glasses were so filthy that I was surprised he could see out of them at all, but his eyes were shining with delight. "Look. Look at what is inscribed inside the wall." I peered at the cup, its fabric turned orange against the sun. Within the thin porcelain was a small section of a chrysanthemum scroll that had been molded into the clay. It was identical to the chrysanthemum scrolls we were finding on other pieces, pieces that were definitely Vietnamese. The fact that fifteen of these eggshell cups were later discovered coiled inside a Vietnamese vase was yet more evidence that they had been made in Vietnam. Other examples were found completely intact, depicting dragons and cranes, while another showed three boys chasing each other through leaves.

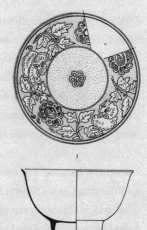

It was a triumph for Ivo. Mensun could not fault the diagnosis, though both he and the local archeologists were at a loss to explain how the Vietnamese had accomplished the feat. Vietnam supposedly lacked the sophisticated refining techniques that removed iron from the iron-rich waters of the Red River delta and made porcelain possible. These pieces told a different story.

Ivo's work at the washing tables was turning up other surprises. Alongside the gaming pieces were fishing weights and hooks, and, from deep in the mud of the bilges, came the bones of countless rats that had not managed to flee the sinking ship. Rats were the scourge of sailing life, spreading disease as they devoured the ship's provisions. Controlling their numbers on a long voyage was vital, and it came as no surprise when later the skull of the ship's cat appeared near the galley at the stern, chasing its quarry to the last.

Some of the things that were being found couldn't be identified immediately. Among these were some fruit stones found inside a jar. They were sent ashore to experts who identified them as coming from the "dragon's eye," the fruit of the longan tree. Longan trees were cultivated in the north of Vietnam, in the Red River delta, where the pottery was known to have been made, so it was no surprise that they were carried on the ship. However, the fruit is extremely perishable and has to have its stone extracted before it can be stored in its dried form. To have had edible longan fruit complete with stones on board, the vessel must have left harbor in August—or July at the earliest—when the trees bear fruit. That was at least two months farther into the typhoon season than we had planned to stay. Perhaps the captain's schedule had slipped for some reason. After all, our own stay was lengthening, postponed so that we might fill another barge with ceramics from the seabed.

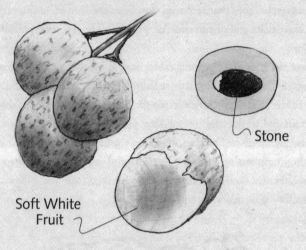

Stone

Soft White Fruit

FOUR WEEKS after the start of operations—which was when the excavation was originally planned to come to an end—the storage barge OL *Star* began its journey up from Singapore, towed by another tug. When she arrived we would have eight vessels on site—the three barges and each of their support tugs, plus the Border Guards' patrol boat and our supply sampan. Working at our current rate, filling the new barge would take at least an additional four weeks, taking our flotilla into July, well into the danger zone.

Ong was not only gambling on the Dragon Sea's temper, however; a great deal was riding on the quality of the ceramics that we were recovering. One day, before my shift, he asked me to meet him in the archeological office on *388* to explain how he felt we should proceed.

"What I think you understand that the others don't is how our two sides must work together," Ong began, once again somehow making me feel that little bit smarter than my colleagues. "Although it can sometimes seem like we're coming from different angles, myself from the business and operational side and you guys from the archeological, we in fact have very similar goals. I've noticed that a lot of the pieces that we are getting up are not of very high quality. They are the basic trade goods. Some of the bowls, for instance, are very poor. The clay's not nice, nor's the painting. Mensun tells me that these are just as interesting for him, but let me put it to you. Which is more valuable, archeologically: twenty identical, poor-quality bowls that have been dashed off by some apprentice, or five good-quality pieces with different, interesting designs executed by a master craftsman?"

"Well, it would depend—" I began, treading carefully.

"My point here is that what works for you can also work for me. If we can recover from an area that gives us good-quality pieces as opposed to lower-quality, then surely we should do that, yes?"

"If everything else was equal. But we can't just jump from area to area cherry-picking," I countered. For a start, we didn't know where to find the high-quality stuff except by excavating. Some of our most beautiful dishes had been packed in among much more mediocre material. There was no way of predicting what would be where. Often the important designs were on the big dishes, and they were stored below the smaller wares in the compartments. To get to them we would need to unpack the smaller pieces, and if we unpack them then we might as well recover them, I reasoned.

Ong nodded, but wouldn't give up. He knew that if he recovered too much, the individual price of each piece would fall; at a certain point it would be uneconomical to get more. There was a danger of flooding the market, and he needed to find a way to prevent that from happening. The desalination tanks on *Abex* already held more ceramics than he had sold at the *Geldermalsen* auction. At the time, Christie's had feared even that was too much. As it turned out, with the buying frenzy the sale generated, they probably could have sold much more. Ong was content to risk filling another storage barge, but he knew that there was no way he would be able to sell the entire cargo that the Hoi An junk had carried. Three-quarters of a million pieces was going too far. Still, he was also concerned about leaving it down there, for fear that collectors and the buying public might find out. The pieces would never hold their value if anyone knew Ong might come back to the site and help himself to more.

In the end, Ong asked me to help him disguise the fact that at least half of the Hoi An cargo would not be recovered. I resented the idea, but as strategies go it was more palatable than dynamiting the unwanted portion—as Hatcher was rumored to have done with the *Geldermalsen* cargo. In the spirit of cooperation, I agreed not to highlight (though I would not hide) the wreck's incomplete excavation on the plans of the wreck site that I was producing.

As we talked, it became evident that other elements of the *Geldermalsen* cargo sale were on Ong's mind. One of its most marketable aspects, of course, had been that pile of gold ingots. Gold was the very symbol of shipwrecked treasure and jumped out of publicity photos. Finding even a little on the Hoi An wreck would add allure, increasing the value of everything else found on it as well as fetching many times its own bullion value. Ong opened the site plan and asked me to take a rough guess as to where on the wreck we'd be most likely to find gold.

The look in his eye was all too familiar to me. Gold is a strange element. Among the most nonreactive of the metals, it remains unchanged on the seabed for centuries while all else around it crumbles and corrodes. It has the same effect on people. Every time I've worked on an excavation where gold has been chanced upon, the project has quickly degenerated. In Mozambique, for example, our experienced, close-knit team began to fragment the moment a cache of contraband ingots was discovered in the shattered bow area of the stricken Portuguese-style *nao*. Trust dissolved, even as the gold at the center of the feuding lay inert, unchanged. The same had happened in the Cape Verde Islands, when a young diver found a gold belt buckle on the wreck of an English East Indiaman, the *Princess Louisa*. He had tried to smuggle the artifact out of the country and been caught as he passed through the airport metal-detectors. High-level strings were pulled to spring him from jail, but the project never recovered. I couldn't help Ong in his request; I had no idea where such valuables might have been carried on a junk, if they had been on board at all. Besides, searching for gold ran counter to all our plans of running a methodical excavation.

Mensun snorted when I told him later about Ong's request. He had different ideas of how we would use our extended time on the wreck, and set his sights on gaining more archeological information—his kind of gold. He had already begun the process, for I had noticed that while his excavation logs were growing increasingly detailed, the quantity of ceramics recovered on his shifts was dropping. As if he had sensed Ong's attempt to push the balance toward more recovery, Mensun was adding weight to the archeological objectives. Now that enough ceramic material had been recovered to rewrite the book on Vietnamese ceramics, he decided that with weeks of extra time on the seabed we should look in greater detail at the vessel and the traders themselves. Initially this had been a secondary goal.

I was torn. On the one hand I had great respect for Ong. Against all the odds, he had mobilized what was shaping up to be the largest underwater excavation ever attempted. Perhaps his methods were sometimes too commercial and profit-oriented, but he was trying his hardest to apply archeology to a business equation. On the other hand I had my loyalty to Mensun and his quest. Though he sometimes seemed blind to the necessities of working within budgets and schedules, he was driven by passion and a higher purpose that made thoughts of economic viability seem pedestrian. I felt as if my feet were in two different boats, their courses diverging.

What none of us knew was that at the other end of the South China Sea Mike Hatcher had managed to pull forty-seven shipping containers of Chinese porcelain from the seabed and had already smuggled it back into Australia. The haul of 350,000 intact pieces was already bigger than anyone imagined a ship would carry, but Hatcher was not finished yet. He would be going back for more, and Ong still had no idea that his rival even had found a wreck.

———

WORKING TO HIS OWN priorities, Mensun decided to use the time and manpower available to him underwater to bring a section of the hull up to the surface for us to examine. Removing it would be destructive, but so is every element of excavation, and at least we would be able to have a good look at it with our own eyes and record it properly with drawings and photographs. Alongside the valuable clues hidden in the tree rings and the cellular structure of the wood itself, a close examination would offer insight into the way the hull had been put together. Mensun chose a section where the bulkhead joined the hull of the ship. The bulkheads were the principal stress-bearing structures. Comprehending these would represent a significant step toward understanding the vessel as a whole.

Ong would never appreciate the point of such a time-consuming operation, said Mensun. The best thing would be to slip the work in when Ong was off the barge, his back turned. On its own it seemed a small deception, but distrust was growing rapidly between the two men. The divide between them was becoming a chasm that threatened everything.

CHAPTER 19

Breaking Point

FATIGUE WAS CREEPING through Kerr's body, gradually numbing him to the danger. He stood on the barren seabed, the beam of light from his helmet searching through the water above, waiting for the empty recovery crate to appear. The crates came down fast and were heavy enough to crush anyone they happened to land on. Staring through the sediment-filled water was deceptive. Shapes would appear when nothing was there, and the other way around. Several of the other divers had said they'd felt something big was down there with them. No one had got a clear sighting, but occasionally Kerr had caught definite movements around the edges of his vision. Whether it was a shark or just a big fish, he couldn't tell. Whatever it was, he was too tired to care, and there was nothing he could do about it. He kept searching above him for the crate. When a dark shadow finally appeared he stepped back, calling for the winch to slow, but the message arrived too late. The crate crashed down in a plume of mud only a few feet in front of him. Kerr swore. That was the way most divers got killed: by getting caught underneath or between heavy objects, or becoming tangled in powerful tools. With saturation diving, you were far from help. Even the smallest mistake could be life-threatening.

Kerr moved over and untied one of the compartment lids. Anchored to the base of the crate, the buoyant baskets inside drifted free and cascaded upward in a vertical line. He pulled the stack toward him, untied the two uppermost baskets, and began to make his way back toward the wreck site. The spot where he was working was once again on the far side of the site. He would have to pick his way across the whole field of grids to get there. Kerr had stopped bothering to protest. He had been at pressure for fifty-three days, sixty if you counted the week before the one-day break. Since the supervisors had cut the length of the bell runs, each shift on the seabed lasted eight hours rather than twelve, but even that was a long time to spend in the water.

Kerr's energy levels were flagging. He had lost weight, and he felt weak. His mind was shifting into survival mode, a dark corner he'd discovered during his initial training for the French Foreign Legion. The outfit prided itself on continuing the battle to the bitter end: When fighting in Mexico in 1863, at the Camerone Hacienda near Puebla, three officers and sixty-two infantry held off two thousand Mexicans for a whole day before the last five Legionnaires still alive fixed bayonets and charged. The name of the encounter still adorns the Legion's flag and infuses its culture, and new recruits were trained with that philosophy in mind. Kerr and his squad were once sent on a fast march across the Algerian desert. When they arrived at the fort that was the finish point, the officer in charge had announced a change of plan and that they must march on to the next fort, many hours away. The physical ordeal had been accompanied by psychological torture. When they finally arrived at the second fort, they were made to stand at attention on the gravel of the main courtyard. The officer then announced that he didn't want to see any more gravel. He wanted it all piled in the corner instead. The recruits were about to begin when he added that if he caught anyone moving more than one stone at a time, there'd be hell to pay. In the

Legion, that meant a beating from the N.C.O. The squad went to work picking up gravel stones one by one. Once they were done (a full day later), the officer inspected the pile. I don't like the way it looks, he said. Put it all back. Stone by stone.

So Kerr knew how to put up mental walls. He was sure that George, with eight years in the SAS behind him as well as years in saturation, would have the same facility. Survival mode was about focusing on small goals, achieving them, and going on to the next. It was when the rhythm was interrupted that things got hard, and that was the problem here. He would be midway through filling a crate, his goals set, when suddenly he'd be told to drop everything. Without warning, he would have to go on a survey, measuring from one point to another apparently at random, occasionally repeating the same measurement three or four times. Sometimes he had to survey for the full eight hours of his dive. He knew the archeologists up top were transferring these measurements onto a site plan, building a picture of the wreck, but that didn't make it any easier. Unable to see how the work was progressing and powerless to suggest alternative methods, Kerr had resigned himself to thinking like a robot, just measuring and loading crates.

MOORED ALONGSIDE *Abex,* the new storage barge, *OL Star,* was filling up rapidly. A week into June, almost twice Ong's original target of 100,000 pieces had been recovered, but trouble was brewing. Not only was a tropical depression lurking to the east of the South China Sea, but the disparate nationalities, professions, and attitudes that had come together for the excavation were beginning to pull themselves apart.

On *Tropical 388,* communication between the teams occurred only when absolutely necessary. In the few waking hours when not on shift, the Indonesian deck hands gathered around the crane near the bow, sheltering in its shade by day or illuminated by its headlights at night. The junior dive techs kicked around the stern. Other crew members sat outside their cabins on the mesh walkway, staring down at any passersby. Another day, another dollar, as Winterberg used to say. Underwater, the gulf between the archeological work and recovery was widening. With the divers working deep in the compartments or in trenches outside the hull, the grid scans were now useless—the silt was so thick that the camera could pick nothing up. All we could do was keep track of what was coming from where. Before, the rhythm had been one of survey then recovery, survey then recovery. Now the surveys were separate activities, done in times when the current swept the water clean or at the start of a bell run when the sediment hadn't yet been kicked up. Though I could see the wreck on the screens and on paper in front of me, I had no physical contact with either the hull or the cargo itself, and the scholarly haven of *Abex* seemed very far away.

Dilip was doing his best to hold everything together. He was trying to accommodate Mensun's needs and requests, but several times a day Ong called him, demanding to know statistics: the number of pieces recovered intact, hours worked on the seabed, location of the divers, hours spent doing survey, and bell-run lengths. Ong also asked him repeatedly if Mensun was interfering with the work. Dilip told him no, and he meant it. If anything, he felt the opposite was true. Mensun always seemed to make a point of asking how he was and telling him that he should take it easy. Everyone's lives were in Dilip's hands, and as Mensun had observed but Ong seemed to ignore, Dilip hadn't had a break for months and needed a rest.

MENSUN SHARED a cabin with Dilip on *388*, and they often had the opportunity to talk. Mensun was genuinely worried about Dilip and the strain he was under, but felt he had no option but to keep pushing the operations manager to cover for his extra survey work. He didn't want a potentially combustive confrontation with Ong; he would rather make sure his own goals were being met discreetly. It would be to Ong's advantage in the long run, he reasoned. The clement weather could not last forever, and he still didn't have anything concrete on the dating of the sunken vessel.

At last, at Ivo's washing table one day, a clue was discovered by Mr. Toan, one of the Border Guards. Toan generally preferred watching the proceedings from a few paces' remove rather than getting his hands dirty. However, he seemed to have a special gift, for whenever he did join in, his fingers would soon alight upon something interesting. His most spectacular discovery to date had been a gold ring, a red stone set in its face. The next time this constabular King Midas touched the mud he found not jewels, but the next best thing: a coin. It was of typical Chinese manufacture, circular with a square hole through its center, stamped with characters. Ivo had a thorough knowledge of the dynasties during which the ship must have sailed and examined it carefully, but was unable to decipher which one had produced it.

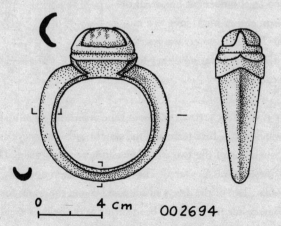

0   4 cm   002694

Over the next few days more coins appeared, turning up in the bottom of cargo holds where they may have accidentally been dropped by crewmen or by those who had been loading the wares. They were sent up to an expert in Hanoi, whose report only increased the confusion. Ivo had been unable to date the marks because the dynasties were so much earlier than expected. The earliest dated back to the seventh-century Tang dynasty, at least seven hundred years before the pottery was thought to have been produced. The most recent dated from 1408, the reign of Yongle. So broad was the range that Mensun began to wonder whether a coin collector had been on board. Who else would have been carrying around a pocketful of coins, some new, some seven hundred years old? On previous wrecks Mensun had excavated, the coins had sometimes been the key to dating the wreck, giving a *terminus post quem*.

The Hanoi numismatist whose report had caused the consternation was the one to clear it up. The range was entirely normal, he assured Ivo over the phone. The succession of dynasties within the Chinese Empire had been so assured that coinage remained constant.  Their coins were recast many times and remained legal tender for hundreds of years—sometimes for as long as a millennium. The coin mystery was resolved. Mensun, however, was no closer to dating the cargo.

ROBBIE HAD BEEN THINKING about time in an entirely more practical sense, watching the dive team and the system with a wary eye and wondering when one of the two would give him a serious scare. Then one day his question was answered. The bell had surfaced normally, just before midday. Two of the divers inside had been in the water for the last eight hours, while the third had been sitting in the bell. As usual, the orange capsule was lifted slowly up to its track and then gently lowered onto the flange above the chambers. Two deck technicians standing on the frame placed the clamps around the junction between the bell and the chamber, then tightened the fittings and stood back. When Sykes tried to pressurize the trunking that now linked the bell to the chamber, the junction leaked. This had occasionally happened before, but this time they couldn't get a seal even after a dozen attempts. Without an air lock to pass through, the men were stuck in the bell.

The divers sat in the capsule, the interior temperature steadily rising in the midday heat. Robbie and Sykes were growing increasingly concerned. It was uncomfortable enough for one man, let alone three. The divers were still dressed in their dive suits; it was too cramped to remove them and there was no Environmental Control Unit in the bell to cool them down. The situation was getting dangerous. Heatstroke could strike. At a body temperature of 41 degrees Celsius (106 degrees Fahrenheit), brain death begins. At 50 degrees Celsius (122 degrees Fahrenheit), muscles become rigid and immediate death is certain. Though the outside temperature on the barge was only about 33 degrees Celsius, the interior humidity combined with the direct sunlight was causing the temperature to mount into the 40s. They'd been through all the procedures and nothing worked. Again and again they tried. Finally, a full three hours later, the seal held and the divers could be decanted. On the monitors, each diver looked drawn as he descended from the bell and collapsed on his bunk. None would admit as much, but Robbie could tell they were exhausted. He would make sure they got an extra few hours' rest.

It wasn't just the divers' bodies that were taking punishment; it was their minds as well. The pressure of saturation leads to chronic mental strain—cabin fever. After about four weeks of confinement, any group will begin acting irrationally. This is one of the main reasons behind the twenty-eight-day saturation limit. NASA undertakes months of interviews and psychometric profiling of its astronauts to select a carefully balanced team for its space missions, yet astronauts on long missions are still afflicted by general nervous tiredness, hypersensitivity, irritability, and listlessness. Divers for commercial saturation are not selected on the basis of their psychological suitability. Get too picky about whom to choose for such high-risk manual labor and you end up with no one. Most divers are not the type of people with whom you'd want to be locked inside a chamber. Many report that the closest parallel to saturation that they can think of is prison, both in terms of the environment and the type of inhabitants.

As the saturation continued, I watched the screens and listened to the communications between the divers, the supervisors, and the life-support technicians. The supervisors, Sykes and Turner, would calmly absorb torrents of abuse as the divers vented their frustrations. They began to treat certain individuals with care, joking with them in an effort to keep their spirits up and to defuse tensions. As always on a long saturation, Robbie was prepared for the siege mentality to develop inside the chambers, but he had not expected it from his own son. Every day Chad became more and more irritated with his father's concerned questions. One day when Robbie called out, "How's it going in there, son?" over the intercom, his boy snapped back, "Stop calling me 'son'! In here I'm Chad, topside."

One night one of the divers on Team 2 was far from the bell, excavating beneath the hull of the ship. The tunnel he was in was narrow and deep, a full eight feet from the entrance; his umbilical was at full stretch. All the supplies flowing down the diver's umbilical—hot water, power, breathing gas, communications—run via the bell, and the bellman would have been very aware of his teammate's position. But as the diver burrowed deeper, he suddenly became aware that breathing was becoming difficult. He cranked the free flow on the side of his helmet to increase the supply. Nothing.

"Topside, I've got no gas! What's up?" he said into the microphone inside his helmet. He heard nothing back. Then he remembered that he was on the dud communications channel; the interference problem still had not been fixed. To speak to the supervisor he was supposed to wave in front of his helmet camera, but in the opaque mud of the tunnel beneath the hull he couldn't even see his hand in front of his face. Instinctively he reached behind his back and cracked the valve of his bailout bottle. The twelve-liter scuba tank, mounted upside-down behind him, carried enough gas for about five minutes at this depth—nowhere near enough time for him to get out of the tunnel, pick his way across the scaffolded site, and make it back to the bell. Backing out of the tunnel, he tried to control his breathing. The rules change when your gas supply becomes finite. Suddenly speed is not everything, and adopting the slow, deliberate efficiency of the scuba diver becomes a matter of life and death.

Emerging at last into clearer water, the diver waved his hand in the light of his helmet torch. Sykes's voice crackled in his ear.

"Hello there, Diver Two, what's up?"

"Lost me gas! I'm on bailout!"

"Okay, hold on." Back in Dive Control, Sykes's hand shot out to the myriad gauges in front of him on the panel, tapping each out of habit as he muttered. "Bell bottom mix okay, reclaim okay, working . . ." He flicked a switch. "Bellman, Diver Two's got no gas and is on bailout. I've got pressure to the bell. What you got?"

On the bell camera monitor, Sykes watched the bellman slowly shift in his seat and inspect the gauges. His hand moved leisurely over the valves, then stopped and twisted one.

"Got it, topside. Valve was out of position. Must've knocked it with me elbow," the bellman squeaked.

"Roger that. Watch it," Sykes replied, keeping his voice level until he had flicked off the switch. "Jesus Christ, what're these silly buggers playing at?!" Then he composed himself again and turned the switch over to the diver.

"Okay, Diver Two, you should have gas back. That a roger?"

There was a slight delay. "Yep."

"Okay, sorry about that, Diver Two. We had a problem up at the bell but we're sorted now. Let's get that bailout replaced and get back to work."

No one would say that what happened was deliberate, but some divers claim that for the bellman to cut off a diver's gas is a common form of harassment. It might seem dangerously short-sighted, for today's bellman is tomorrow's diver and there would be little doubt as to who had been responsible, but after weeks in saturation, logic takes a backseat.

BY MID-JUNE the strain was at breaking point, above as well as below the water. The sea was millpond calm, and had been so for ten days, but unseen currents under the surface were swirling at greater speeds. The South China Sea's tides surged through the islands of Indonesia and the Philippines in their relentless pursuit of the moon's gravitational pull, gathering power as the spring tide (when both the sun and the moon pull the ocean from the same direction) grew nearer.

The threat of typhoons hung in the air. We were now deep in the season, and we knew how fast a storm could hit. The crews were worn out and fractious. Originally scheduled to have gone home more than two weeks earlier, they saw no end in sight. Our expectations had already been far surpassed in terms of the cargo we had recovered. All our original estimations were based on recovering 100,000 artifacts, and in the storage tanks on *Abex* and *OL Star* we now had more than twice that figure. Dilip, his hair now distinctly speckled with gray, sensed the mood and pushed Ong to set a new end-date on which to finish operations. Ong refused.

It seemed to me that Ong was worried that his lieutenant had been exposed to Mensun for too long and that his loyalty was waning. With his constant shuttling between Hanoi, Saigon, Singapore, Kuala Lumpur, London, and New York, trying to keep the money flowing and preparing for the next stage of the project, Ong had not been able to spend much time on the barges. He wanted to rein Mensun in, but his only chance to do so was by keeping Dilip on high alert. Dilip, however, was getting harder to reach. Though Dilip always blamed this on bad weather and interference with satellite phone signals, Ong must have suspected that his operations manager was avoiding his calls. He couldn't see that the constant pressure was pushing Dilip to the edge.

Ong had the funds to continue and the space in which to store more artifacts. He wasn't inclined to follow Dilip's suggestions, especially now that his operation manager's loyalty was suspect. On the *Geldermalsen,* the chest of gold had been discovered in the last few days of the salvage. Ong therefore decided to hold on for as long as possible. He couldn't know that every day the expedition remained on site was bringing him closer to disaster.

CHAPTER 20

An Ominous Wind

THERE WAS ANOTHER reason to remain on site. Over the past few years, Mensun had been starring as the presenter of the television series *Lost Ships* for the Discovery network. Mensun and the series director, Matthew Wortman, thought the story of the Hoi An wreck would make a perfect episode. Nonetheless, they were having trouble getting the channel to commit the funds. They decided to approach Ong. A documentary on the excavation would provide invaluable publicity for his auction, publicity for which he might be prepared to pay. Ong was interested, though he knew that for a program on the Hoi An wreck to be of any value, the timing had to be right. It had to be broadcast not more than a few weeks ahead of the auction, and certainly not afterward. At this stage, he did not know exactly when the sale was going to be held. The only way Discovery would allow him any say in the timing of the broadcast was if he provided a full third of the production budget, almost $100,000. Ong weighed this proposal carefully, and eventually decided that the publicity generated by a documentary would probably be worth the expense.

Both the barges were busy with activity when the three-man film crew arrived. After two days of shooting, Matthew Wortman had filmed all he could of the work in progress. Now he needed to build up the drama and develop the story. For a show to do well, it had to feature lots of action and a gasp every few seconds. To catch actual drama, the camera crew would have to have been on board for the whole project, and that would be far too expensive. Instead, they would have to re-create a few scenes. Fortunately, with three episodes under his belt already, Mensun was aware of the need for enthusiasm and vigor on camera.

Wortman learned that there had been a rash of pirate attacks one hundred miles south of the wreck site, and he decided that a fictional brush with the brigands might make for an exciting scene. He persuaded Ong to don some white expedition overalls and join Mensun on the bridge of *Ena Supply*. (Being without engines, *Tropical 388* didn't have a photogenic enough command post at which the pirate warning could be received.) Wortman announced that they would do a few takes unscripted and see what happened.

"Action!" Wortman yelled. Ong blinked in surprise as Mensun jumped forward and grasped the wheel.

"ALERT ALL CREWS! WE'VE GOT A RED-ALERT PIRATE WARNING!" Mensun shouted, as if to some waiting executive officer, and dived onto the flying bridge outside. He scanned the horizon with his hand, shading his eyes, and then leapt down the stairs and out of shot. Wortman continued panning to reveal Ong, standing by the side window, hands at his sides, his mouth open.

During the course of the week that the TV crew was filming, Ong grew more and more exasperated. Mensun had pitched the idea of the documentary to him as a fantastic marketing opportunity. So far it had seemed to be more about publicizing Mensun than about the wreck and its cargo. Ong had never wanted to be a treasure-hunting celebrity like

Hatcher; it wasn't his style. All he wanted was to orchestrate a successful and profitable project. But Mensun was now monopolizing the spotlight on a worldwide TV show, and claiming the wreck as his own.

MENSUN WAS back at work analyzing the cargo—the pirate attack successfully repulsed—when Ivo appeared.

"Mensun! Look at this!" he exclaimed, holding out a medium-sized blue-and-white dish. Within its border of lotus-petal panels, the dish featured a painting of a tree, two concentric circles beneath its canopy. Mensun had never seen anything like it. This was not unusual, however; new designs were appearing every day.

"What are you seeing, Ivo?"

"This . . . this is not a conventional design."

Mensun nodded, encouraging him to go on.

"The strange thing is the sun. If you see, it is in fact two concentric circles with an inscription inside," continued Ivo, pointing to the disc beneath the arc of the tree. "They are Chinese characters. They say 'Moon' and 'Sun.' In that order. Usually in Chinese the sun always comes first. It is bigger, and more important. In all traditional inscriptions the sun comes first. But here it doesn't. So why? Well . . ."

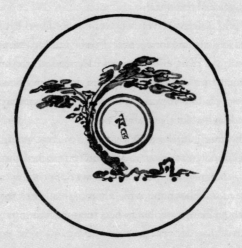

"An eclipse?" supplied Mensun, who had taken the dish and was hold-ing it with both hands at arm's length. "You're right, Ivo. It does look that way."

"And eclipses were very important events, very auspicious. They were always recorded. But if an artist has chosen to paint one, one must have occurred in recent memory." Mensun was nodding now as Ivo spoke.

"I see where you're coming from, Ivo. It's worth a shot. Where are you going to find records of when eclipses were visible over northern Vietnam?"

"The *Dai Viet*," Ivo replied. "The ancient chronicles of the North Vietnamese kings. They have just been republished, and they would record all such events."

Mensun wasn't sure that a string of suppositions would be enough to pin down a date for the wreck, but he was once again thankful to have Ivo on board.

Dilip's cell phone rang, the tone lancing through his headache. He cursed, and walked up the stairs to the walkway where, to his avowed re-gret, he usually managed to get a network signal. Ong had been calling re-lentlessly and Dilip occasionally answered, forced by his sense of duty. He needed a break. He hadn't had a day off for months, and he could feel the fatigue affecting his judgment. His birthday was coming up, and Mensun had suggested he go ashore for a couple of days. Things were pretty much running themselves right now, Mensun had pointed out, and fighting fatigue was counterproductive. What Mensun said made sense to Dilip, and he had begun to look forward to a small respite. It would be just long enough to refresh him for the last few weeks of the job, which Ong now wanted to carry into July. Given the way Ong kept postponing the depar-ture, the barges would leave only when chased by a storm. It would be an-other dramatic last-minute exit, and Dilip would need to be on his toes.

He answered the phone and replied mechanically to Ong's string of questions—about how many hours had been clocked, how much had been recovered that day, and so on. The figures were down, Dilip explained, because of the currents: They were running hard at midtide and risked pushing the bell underneath the barge, making it impossible to bring it back on board. Dilip had experienced that situation himself on a job; he had had to wait six hours before the deck crew had been able to bounce the capsule off the hull and over the lip. If there were any doubts, you didn't dive.

But Ong was asking pointedly whether the shortfall was because Mensun had been interfering again, pulling the divers off recovery duties to do survey work. Admittedly, surveys had been done, but as far as Dilip was concerned, that had been the original deal. Anyway, the point here was the currents, not what Mensun was doing. Ong ignored him, insisting that he keep a close eye on Mensun.

Dilip felt a surge of rage build up inside him. The directors' disagreement was becoming nightmarish. Did Ong not remember that Dilip, too, had invested his own contacts, time, and money in this project? He wanted it to succeed as badly as anyone. Did Ong not trust him to do the job? Dilip's loyalty was already cracked but now anger began streaming into his blood. He'd had enough. He took the phone away from his ear and pushed the "end call" button so hard that something splintered inside the handset. Ong's voice was cut off mid-sentence.

"Fuck you," Dilip said, through clenched teeth.

LATE ONE NIGHT Chad Robinson was working at the stern of the wreck. I had been watching him on the monitor as he exposed a large concretion of iron pans where we supposed the ship's galley to have been located. We were nearing the end of the bell run and the current was picking up. Chad was forced to brace himself against the grid supports in order to stay in position, but the flow of water meant I had a clear picture on the screen. He was taking handfuls of mud, breaking them with his gloved fingers to check that nothing was inside them, and feeding them to the airlift.

Chad slowed, his rhythm interrupted. His hand reached up and turned off the flow to the airlift. The image jumped around for a couple of seconds as the airlift bucked, and then sank to the ground. When it steadied, the picture refocused on a patch of mud.

"Can we see what, Diver One?" I heard Turner ask.

The camera panned down, then refocused again. In the center of the screen, half buried in mud, a skull stared outward. Chad's hand gently waved more mud away from it and a jawbone appeared, followed by other bones.

Skeletons on shipwrecks are rare. Jacques Cousteau made this observation in *The Silent World,* noting that in all of his dives (some five hundred), he and his team had almost never found human remains. He observed that "very few victims drown inside sinking ships. They get off beforehand and drown in the sea." Popular imagination fills the coral-encrusted, weed-strung timbers of a shipwreck with leering skulls, but the reality is that when a ship goes down fast, a person's instinct to get off is usually faster, and once in the water, most bodies float away. The older the boat, the less likely you are to find bodies. Older boats tended to have fewer decks, and were generally smaller. Many souls might perish in a shipwreck, but it is unlikely that many skeletons will remain on board.

There are exceptions, of course. In 1790, two Turkish salvage divers anchored above a wreck at Senaglio Point in the Bosporus. The water was not deep, and the wood-and-leather pump they kept on board their boat would manage to deliver enough air to their leather headdresses. They were hoping for treasure, but minutes after being lowered they started signaling frantically to be pulled back up. One hundred and fifty years earlier, the Ottoman sultan Ibrahim "The Mad" (1616–1648) had received a tip-off that one of the girls in his 280-strong harem had slept with another man. Sultan Ibrahim was not a man of normal passions. Dimitri Cantemir of Moldavia wrote: "In the palace gardens he frequently assembled all the virgins, made them strip themselves naked, and neighing like a stallion ran amongst them and as it were ravished one or the other, kicking or struggling by his order." When the sultan heard news of the infidelity he had reacted with characteristic intensity. The girl who had informed him was his favorite concubine, Sechir Para ("Sugar

Cube"), who at about 330 pounds was also the heaviest in his harem. She was spared his punishment, but the other 279 women were sewn into stone-weighted sacks with only their heads protruding, then lashed to the deck of a boat that was sent to the seabed. The divers had approached the shipwreck through the gloom only to be greeted by "hundreds of bowling-pin shaped forms with grinning, lipless mouths swaying in the current."

Looking at the skull, which was half buried in mud, on the screen reminded me of when I'd first encountered a skeleton underwater. We were searching in ever-widening circles through the opaque murk of Maldonado Bay in Uruguay, looking for the remains of the Spanish troopship *San Salvador,* which had sunk only five hundred feet from shore in 1813 (four years after the *Agamemnon,* which lay a quarter mile farther out in the bay). Feeling my way across the seabed, my hand happened upon what felt like a large cannonball. As I peered closer in the green water, inches in front of my face the cannonball finally resolved into a skull. Though the ship sank close to the beach, more than six hundred troops, women, and children had died on board. Very few people knew how to swim in those days; for a sailor, to learn was to invite bad luck. We came across the skeletons of a man, a woman, and a child near the main-mast, surrounded by medical instruments. The finds confirmed what eyewitnesses had seen from shore: The doctor and his wife and child had been trying to get to the lifeboat when the mainmast had fallen, trapping them in its rigging as the ship sank. The only survivors had been the captain, the pilot, and the priest, who escaped in the single lifeboat.

Though they can yield valuable information, human remains on a wreck stir up not only emotions but legal issues, which can easily cause projects to be shut down. In this case we'd been warned that any bones should be left on the seabed unless we obtained special permission from the excavation committee. Once Chad had taken some reference photos, he was moved to another grid.

THE NEXT DAY Mensun summoned me to *Abex,* refusing to explain the reason over the radio. I arrived to find him furious. After the arrival of the latest crate, all hell had broken loose. When they had raised one of the compartment lids the skull had appeared, carefully placed in the uppermost layer. Human bones are the most sacred of all objects to the Vietnamese, who believe that touching them brings a lifetime of bad luck. Disturbing such remains from their resting place is thought to attract a similar curse. Upon the discovery of the skull, the washing crews had refused to go near the crate; work had ground to a halt. I was at a loss, unable to explain how the skull had got there, beyond suggesting that one of the divers had decided to play a practical joke.

Hoang Van Loc from VISAL quickly took control and rushed a priest out from the mainland to conduct an exorcism. Only after the entire crew had solemnly paid their respects to the skull, amid swirling incense and incantations, did the teams finally go back to work. For the rest of the excavation the skull remained in a desalination tank of its own, at the very end of *OL Star.* The washing crews gave the corner tank the widest berth they could manage.

Despite the uproar it had caused, the skull eventually provided an important clue when it was analyzed. Professor Nguyen Lan Cuong from the Institute of Archeology in Hanoi compared the features of the cranium to a database of attributes and declared that it bore the closest resemblance to a young woman, age eighteen or nineteen, of Thai origin. A copper ring had been found near one of her arms. A rumor spread that she had been chained to the deck, a prisoner. Mensun scoffed at the idea, but when altarpieces and more female bones were discovered nearby, he privately wondered whether the other women on board had been praying when their end had finally come.

A STIFLING, oily calm had now surrounded us for weeks, the surface of the sea disturbed only by the currents that tugged at the barge, boiling up around the hull. One night, soon after the unexpected appearance of the girl's skull, an unusually warm breeze blew up. It was nearly eleven, and I was due to start my shift in Dive Control an hour later. I wrapped a towel around my waist to walk to the shower and opened my cabin door to find the air thick with movement. The deck lights were reflecting off the carapaces of thousands and thousands of insects. We were twenty miles from land, but they were everywhere, crunching under my feet on the walkway and coating the equipment and decks below. Shielding my face, I pushed my way into the shower. The striplights were shrouded in buzzing, crawling black forms. As they fell around me in the shower, senseless in their swarms, I saw locusts, cockroaches, and countless other types of flying beetles. They bounced madly between me and the walls, getting caught in the jets of water and sent sprawling over my head and shoulders.

DILIP'S BIRTHDAY—and his mini-break—was fast approaching. He was holding on tightly to the prospect. As we talked one morning he dropped the smoking butt of his cigarette to the deck and stamped on it, catching a beetle with a crunch as he did so. Without pausing for thought, he took another cigarette from the pack in his pocket. The stress was unlike anything he had experienced in saturation, where he could switch into automatic and take the minutes one by one. His eyes were now permanently narrowed and his teeth gritted, the front two turned brown from chain smoking. The responsibility of command—at least the responsibility of *this* command—was crushing. Any normal project would be fine, with concrete goals set and everyone playing for the same team. Here, Mensun was trying to do one thing and Ong another. Mensun seemed to be going about things in an honorable way; Ong was constantly suspicious of his partner and always on the phone checking up on him. When Dilip informed him out of courtesy that he was going to go ashore for a day, Ong had seemed very twitchy. That couldn't be helped, said Dilip. He needed the break, full stop.

Ong arrived on board by surprise on the day before Dilip was due to leave. Within minutes they had disappeared into the barge office, and closed the door behind them. Dilip later told me that Ong had ordered him to stay at his post because if he left he would give Mensun too much control.

"Listen, just out of curiosity, why get so worked up?" Dilip had asked.

"You don't know Mensun," Ong reportedly replied. "Mensun will do something to harm the project if you leave."

"Ong, come on. What can he do to jeopardize the work?"

"You don't know Mensun that well. I know him. The moment you are off he will start his rubbish."

"What rubbish? Tell me now."

Ong could give Dilip no convincing answer. An argument broke out. Finally, Dilip snapped. He was being told to cancel his weekend away out of an unfounded, petty suspicion. He was being ordered to put the safety of the project on the line because of some personal disagreement. Fury boiled up inside him. If success came at a price, that price was too high. He reined in his voice and yoked it to what authority remained in his grasp. Even if Ong was the ultimate boss, Dilip was in charge of these barges. This was his jurisdiction.

"You have outstayed your welcome on board this barge, Mr. Ong," he declared, his voice steely with barely suppressed rage. "Please leave."

Ong stared at him, eyes beginning to bulge, but Dilip continued. "As long as I'm running this project I have all the rights. You might be the guy paying the bills, but I'm responsible for this barge and you have outstayed your welcome."

Ong remained seated for five minutes, stunned, then stood up. He had no choice. He picked up his bag and left the barge.

Dilip retreated to his quarters. He had had enough. He no longer gave a damn about the shared future he had envisaged with Ong, and the trust they'd enjoyed was gone. His contract was almost up anyway; he would write off the investment that Ong had persuaded him to make in the project. If this was life outside the oil sector, he didn't want it. He decided to go back underwater, where the challenges were real, not bound up with ego and subterfuge. At forty-two he was becoming an elder statesman among divers, but his experience would be highly valued on long saturation jobs. He disappeared from the barge without fanfare or farewells. Everyone sympathized. They'd seen his gradual disintegration over the past few weeks as his energies were torn apart by the differences between Ong and Mensun. But his departure was like losing a keystone from an arch; though the void was rapidly closed, the project was now dangerously unstable.

BEFORE HE LEFT, Dilip handed over all responsibility and documentation to Winterberg. The already fragmented atmosphere on board 388 worsened substantially. As a Singaporean, Dilip had been able to straddle the cultures of East and West, smoothing the hard edges of Western manners for Southeast Asian sensibilities. Winterberg's abrupt Germanic manner did the opposite, and the Indonesian deck crew soon began to resent their barge master's orders. Late one night, as I walked around the perimeter of the barge, I heard raised voices. Near the crane, Winterberg and the crane driver were nose to nose, snarling. Winterberg towered over the Indonesian, but by the time I reached them the confrontation was over and Winterberg was steaming back sternward.

"The little fucker. Let him try."

"What's going on, Hart?" I asked when I caught up with him.

"He says he's going to kill me. So I say, name your time."

"What? Why? What happened?"

"Listen, it's no problem. No worries. Everybody says they want to kill me. I said yes, just go ahead and try. But do it good because you won't get a second chance, you know. No problem, I said. Everyone is welcome to try. As you can see, I'm still alive."

ONG RETURNED to the barge a week later. He called a meeting on *388* between Winterberg, Mensun, and me, to decide how much longer we should continue. Though the weather remained good, bell runs were being cut short, and not because of crew fatigue. The currents were running hard for most of the day, tearing at the bell as it entered and exited the water and making the supervisors nervous. A long, low swell was building, too, running from the northeast. The Dragon Sea was emerging from its hibernation. Nonetheless, Ong wanted to stay on site for another two weeks.

"The funds are secure," Ong insisted. "We still have storage space on *OL Star,* and there are big areas of the wreck we haven't touched."

"You can never finish a wreck, Ong," Mensun replied. "Nor should we feel we should try. If someone comes up with a new technique in the future, they've got to have some untouched areas to work with. So it's good practice to leave areas as they were."

"What, for the fishermen? Come on, we're not a charity."

"We have more than completed our original archeological objectives," Mensun replied, his voice stiffening. "When the information we're getting is no longer worth the risk, that's when we should pull down the curtain."

"This is not just about information, Mensun. There may be valuables that we've missed!" Ong immediately fired back.

The debate continued, each side becoming more adamant. I watched Winterberg's downturned eyes begin to twitch as he listened to the two go back and forth. At last it was too much for him.

"Listen!" he interrupted loudly. "You can't mess with the sea. It's no joke. When these typhoons blow, they blow hard. That's fine. I like a good storm, you know? But you, you should think carefully, Ong. When they come you maybe can't move for months. Not move one bit. So that's your treasure stuck in the bay there, in the middle of islands in the middle of sea. And that's this barge stuck too, with all of this equipment. That'll cost you real money."

He had all of our attention now. As he warmed to his theme, Winterberg's voice began to boom. "And I tell you something else, I'll sail *388* back to Singapore but the slightest sign of trouble and I'll cut her loose, just like that. I will not be responsible for her. So, you must plan, you must not get greedy. So. You want decision? I give you decision. I GIVE YOU DECISION! WE LEAVE NOW!"

Both Ong and Mensun blinked at this outburst from the bearlike, white-haired Austrian. Ong looked at both Mensun and me. Winterberg's speech made sense. Dive times were already being cut short, and the weather wasn't going to get any better. The decision was made to leave. Though less than one-third of the compartments on the wreck had been excavated, more than 250,000 intact pots were stored aboard

the decks of *Abex* and *OL Star*. The remainder, still packed within the
compartments and so protected from the fishermen's rakes by the junk's
thick hull, would be left on the seabed. I was stunned. It wasn't at all the
ending I was expecting. It seemed unnaturally controlled. After the skull
had turned up on deck, somehow all of the superstitions that had surrounded our barges for so long seemed to start taking physical form,
from the insect inundation to Dilip's sudden departure. But the bad
omen hadn't been fulfilled—we were leaving the Dragon Sea of our own
accord, by voluntary decision. I'd imagined we would be chased from the
site by a typhoon, or at least somehow forced to leave—perhaps due to
a confrontation between Ong and Mensun. To cap it all, we were leaving with our cargo intact. After a final two bell runs to tidy the site and
pick up equipment, the anchors were recovered one by one and we
calmly slipped away.

CHAPTER 21

Taking Stock

TROPICAL 388 joined *Abex* and *OL Star* in Cu Lao Cham Bay, the three barges dwarfing the fishing sampans and the scattered huts of the village. The divers began their decompression. After fifty-nine days in saturation—almost twice the regulation limit—the six divers were brought back to surface pressure. It took sixty-two hours for the accumulated gases to seep from their tissues. When they finally emerged through the chambers' deck hatch, all were thin and bearded, their skin yellowed and covered in rashes and lesions. Robbie watched his son emerge. He hadn't wanted to see Chad go inside in the first place, but now that it was over he was full of pride for his pale, bedraggled son. He gave him a hug and offered him a cigarette, which Chad accepted with a shaky grin.

I'd expected the divers to hit the whisky immediately, but they couldn't drink for another twenty-four hours. They were on a "bend watch," to make sure their bodies had not saved one last bubble for the bloodstream. Instead of partying, Kerr, George, Chad, and the others wandered the deck like ghosts. The crews gave these pale strangers a wide berth.

Winterberg left with *Tropical 388* as soon as he could, running southward in an attempt to make Singapore before the weather closed in. Previously oblivious to the threat of storms, Ong became terrified that a sudden typhoon might return his treasure to the seabed. The island bay sheltered *Abex* and *OL Star* for the time being, but he was increasingly aware of its shortcomings as a safe haven. He had hardly noticed the tiny, high-walled concrete storm shelter on the main village's seafront before; now it seemed ominous. The bay might be the best refuge in the area, but evidently the fishermen still felt the need for a panic room. And it wasn't big enough for even one of Ong's barges.

With the junk's cargo successfully recovered, Ong's challenges had only begun. He needed to organize the ceramics in such a fashion that they could be divided up between his company (Saga Horizon), the Vietnamese state salvage agency VISAL, and the National History Museum in Hanoi. The museum would retain all the artifacts that were unique to the cargo or belonged to the ship itself, and a representative selection of everything else. The remaining pottery would be split between the partners: Saga would get 40 percent, VISAL 30 percent, and the Vietnamese National History Museum 30 percent (which they would then divide among the other national museums).

Only after the division had been completed could Ong begin to plan the auction in detail and start arranging to take his share out of the country. He had at last succeeded, as he put it, in "turning his investors' funds into clay"; now he needed to turn the clay back into money. To do this, he would have to rely heavily on Mensun. In fact, he was all too conscious that without the archeological director's stamp of approval, he would not be allowed to export the pottery at all.

In this respect Mensun had some leverage over Ong, but then so did Ong over Mensun, for the archeologist's work was also only half done. He had all the raw information from the seabed, but his analysis of the ceramics—the core of his reason for undertaking the excavation—was barely begun. He needed Ong's funding for the post-excavation work and the publication that he hoped would be the foundation for all future studies in Vietnamese ceramics. To have participated in such a controversial project and not publish first-class information would ruin his reputation.

Originally, the desalination phase of the post-excavation was to be carried out in Cu Lao Cham Bay. However, we were now late in the season, and the bay was no longer safe. Deciding where to move the vessels and their precious cargo was by no means easy. Relocating the pottery onto land would be an enormous task. It needed to be kept wet, and the black desalination tanks weighed over two tons apiece when full of water. To make even a short journey their contents would have to be well padded, and that would mean unpacking and repacking each container.

There were political concerns, too. So far, Ong had been dealing with the Quang Nam provincial government. He knew that if the cargo crossed the line into another province he would have to pay crippling taxes. This had nearly happened when he had sent *Abex* to Da Nang to fill her tanks with fresh water for the desalination. At the last minute, VISAL warned him that Da Nang was in a different province and that by taking the pottery across the border, he would be liable to pay a percentage of the cargo's value to the Da Nang provincial authority. *Abex* had turned back, and *388* had gone in her stead.

The Hoi An project had entered a dangerous phase. The piracy warnings that continued to arrive on the Weatherfax were of some concern to Ong, but a greater worry was the buccaneers in suits. Some of the most impressive salvages ever performed had become financial disasters as a result of legal and governmental predators swooping down on the salvager once he came ashore with his prize. In the early 1980s, for example, Keith Jessop spent two years with saturation divers eight hundred feet down in the freezing Barents Sea north of Norway, recovering Russian gold from the World War II wreck HMS *Edinburgh*. Jessop had done everything by the book. His consortium was to keep 45 percent, with the Russian and the British governments sharing the remaining 55 percent to the tune of two-thirds for the Russians and one-third for the British (exactly the way the insurance had been split between the owner and the carrier at the time of its sinking). The salvage operation was an unqualified success—all but thirty-four of the 465 gold bars were recovered, valued at the time at about $100 million. As soon as Jessop landed in Britain, however, he was ambushed by the tax office, which imposed crippling import and capital gains taxes. After several years of costly legal disputes, Jessop's $3 million share had been whittled down to less than $30,000.

It wasn't a new story. Writing more than five hundred years before Christ, Herodotus told the earliest cautionary tale of treasure-hunting contracts not being honored. Fifty years earlier, a diver named Scyllias had made a deal with Xerxes, king of Persia, to recover "an immense treasure" from some Persian galleys that had sunk in a battle with the Greeks. Scyllias managed to recover the treasure with the help of his daughter, but the king then refused to hand over their reward. In this case the salvagers took revenge; a few nights later during a storm, they slipped from the king's galley and cut the anchor ropes of his entire fleet. Xerxes's armada was destroyed, and Scyllias and his daughter escaped by swimming nine miles to shore.

Wanting to preclude any such problems, Ong had tried hard to make sure that his agreements were watertight and that he could trust the Vietnamese committees. He was almost certain that he'd managed. They had an interest in his succeeding financially, after all. VISAL was a state-owned company and had a stake in the outcome of any eventual sale. But "almost" was not quite good enough. Before he moved anything ashore, Ong wanted to be absolutely certain that he could get his cargo out of the country. While it was still on his barges he had a certain amount of leverage; once it was on land, exporting it could become much more difficult.

One day, as Ong flew up from Ho Chi Minh City to Da Nang, he saw a horseshoe-shaped inlet on the coastline beneath the plane through a break in the clouds: Ky Ha. It was the perfect solution: The sheltered bay was just inside Quang Nam province, and the pottery could remain on the barges. The archeologists could continue living on board, working exactly as they had been doing in the bay in Cu Lao Cham. No other accommodation or eating arrangements would have to be made. Ky Ha was well protected, too, being both a natural harbor and the coastal Border Guard's regional base.

KY HA BORE THE SCARS of recent history. The peninsula that formed one side of the inlet had once been home to the U.S. air base Chu Lai, an adjunct to a much larger facility at Da Nang. Chu Lai had been the focus of the first major ground battle of the Vietnam War, Operation Starlight. U.S. intelligence officers had discovered fifteen hundred Vietcong gathered in preparation for an assault on the airfield, and on August 18, 1965, U.S. Marines launched a preemptive strike against them. Though forty-five Marines were killed and 120 were wounded, they took 614 Vietnamese with them as they routed the force. Such a decisive victory at the beginning of their ground campaign sent the morale of the American troops sky-high. Ten years later, in March 1975, their spirits couldn't have been lower as they retreated from Chu Lai, one of the last airfields to be evacuated. When the American forces finally pulled out, the whole base was flattened and the runways were bombed to prevent their being used by the Vietcong. All that remained a quarter of a century later were craters and the fragmented outlines of foundations.

Industry had since taken over the bay's beaches. A few hundred feet away from where our two barges were moored, an enormous steel ship lay like a beached whale, pinned against the shore by a web of lines. Its hull had turned a deep red-black from rust and lacked any sort of superstructure. All day long, daggers of flame were visible through the vessel's sides as torches cut through the steel, strips of which were gradually torn away like metallic blubber.

As fast as our neighbors were pulling apart one ship, we were putting another back together. After the disorientating shift work of the excavation phase and the dissociation between *Abex* and *388*—the archeology and the salvage—work in Ky Ha was steady and pleasurable. At last I was getting a chance to look at the pottery properly. The designs engrossed me. I found myself getting lost in them: spectacular mountain scenes, lions chasing a flaming ball around a vase, or fishermen crossing mist-covered lakes headed for distant pagodas.

At the end of the day, Mr. Hy the artist, Magnus, and I would sometimes swim to a raft in the middle of the inlet on which sat a tiny hut. The resident fisherman would blow on his coals and cook us a fish he had pulled from the nets that surrounded his floating home. It felt like swimming into the decorative scenery on the ceramics, scenery that might as well not have changed for five hundred years.

Ky Ha suited Mensun as well as Ong. The bay's isolation meant no distractions. Mensun needed his team to be working flat-out if he was going to get his publication ready in time for the auction, as he had agreed with Ong. Still, despite the work ahead, he felt himself relax just a little. The nagging fear that his cargo might be taken away from him was lessening. It was off the seabed, safely stowed on deck, and every day he was assimilating more information about it.

Mensun and I spent the days cataloging the wares, describing the different shapes and decorations, and directing the draftsmen while Magnus photographed endless pieces, individually, in groups, and in close-up detail. He had built a studio on board in which he and his two assistants photographed almost four hundred artifacts a day. Even at that rate, capturing a single image of every piece in the cargo we'd recovered would take two and a half years. Therefore Mensun selected representative examples, extremes of categories, and pieces revealing pure artistic merit— as well as the pieces he had mentioned in his text. For each item he had to specify which view he wanted, and whether he wanted a particular detail or the whole artifact.

Photographs are an essential reference. The archeologist's main tool, however, is descriptive technical drawing. A good draftsman can pick out the important details and present them with a clarity that no photograph can match. As on a well-designed map, the important information gets accentuated. It was a lesson I felt more keenly after working with the revolutionary time-saving technology I'd sent to the seabed: my grid-scanning camera dolly. Early on it had become obvious that even in clear water, the browns of the ship's timber and the brown of the mud surrounding it were too similar to distinguish from one another. Other wreck photomosaics that I'd built had taken a lot of preparation—cleaning the wood, lighting, and waiting for the right water conditions. Trying to use the scanning grid when divers were working, what little information the images carried disappeared in a cloud of silt. We'd rapidly returned to the tried-and-trusted techniques of drawing maps.

Mensun had set ideas as to how he liked artifact drawings to be presented. He supplied the eight draftsmen with selected items to work on, then checked the drawings, suggested revisions, and checked them again, his critical eye ensuring that everything the draftsmen produced was both accurate and attractive. He enjoyed working with the artists, and though they were all paid professionals he responded to each in a different way, sensitive to their individual needs and gifts. Mr. Hy, for instance, was an artist, not a draftsman. He quickly grew impatient with the precise replication of line and wash that gave such pleasure to the others. Mensun therefore offered him a different challenge: to reconstruct paintings that had deteriorated over the centuries.

While most of the Hoi An pieces had been painted with cobalt blue lines on the pale clay background and then glazed, some had been given a second layer of detail in other colors. This "overglaze," or enameling, was one of the aspects of the cargo that had most excited the connoisseurs of Vietnamese ceramics. Red, red-brown, yellow, blue, dark green, and white could all be produced, but the technique of applying enamel was complicated, requiring a second firing of about 700 degrees centigrade, hot enough to fix the enamels but not so hot as to melt the first layer of glaze that had been fired at about 1300 degrees. Now, as then, because of their complex manufacture and the extra artistic input, the polychromed wares were more valuable than the simple blue-and-whites. Indeed, the most expensive Vietnamese pot ever sold was a polychrome vase with thick gold leaf highlights auctioned by Sotheby's for £350,000 ($600,000) in November 1999. At the time, there was some debate about whether the gold was authentic but some of the enameled pieces in the Hoi An cargo still bore scraps of goldlike luster, suggesting that the gilding was original.

Sadly, none of the enameled pieces on the wreck had survived in pristine condition. Over the centuries of immersion, the second layer of glaze had slowly reacted with the seawater and was now little more than a collection of misshapen red blobs, green stains, and gold flecks. It seemed as though the original designs on this second layer had been lost forever. Then, one day, when evening sunlight was slanting through the draftsmen's window, one of the enameled dishes was caught at an oblique angle by the light, and Mensun noticed that the designs on the outer layer had left a ghostly matte shadow on the gloss surface of the glaze. He realized it might be possible to re-create what the enameled pieces had looked like before their detail disappeared. He asked Mr. Hy if he wanted to try his hand and he leapt at the opportunity. The only accurate way to transcribe the designs was to pencil in the ghostings on the pot itself until they were dark enough to trace. Mensun was squeamish about doing this, for interfering with the original artifact in such a way would not be regarded as good archeology. For the sake of recovering the original design, however, he decided to run the risk.

It paid off. Mr. Hy eventually presented a complete rendering of what Mensun would later pick out as his favorite piece in the entire Hoi An cargo, dish number 2735. For the Vietnamese, as for the Chinese, the depiction of mountain scenes was an art practiced with the same reverence as calligraphy. The sparse blue-and-white landscape on 2735 had previously featured only hills, a pagoda, and a lotus pond, but after Mr. Hy's reconstruction the landscape blossomed with deer, lakes, wooded hills, and lotuses in various stages of bloom. All this had previously been invisible to the naked eye, as had the two peacocks that now brought the whole glorious scene to life, filling the sky with their fanned tails.

Enameling had begun appearing on Chinese porcelains during the Xuande reign (1426–1435) but the method would have been a closely guarded secret. What intrigued Mensun was that if his theory about the cargo being from the heart of the golden age, in the early to mid-fifteenth century, was correct, then the Vietnamese appeared to have developed enameling around the same time as the Chinese. The established view was that the Vietnamese had picked up the technique from their Chinese neighbors. What if the Vietnamese had developed it independently, or even come up with it first?

———

ONG ARRIVED on the barges one afternoon in late September and began pacing the decks. While Mensun's mind was absorbed by dates that had passed more than four hundred years before, Ong was thinking about the near future. By this point he had got enough assurances from the various governmental committees that the export permits for the pottery would be granted, pending Mensun's report. He was impatient to see the conservation process end so that he could move on to the next stage in his schedule: splitting the cargo among the various parties and beginning pre-sale preparations. A major reason for his impatience: He'd just heard news from Singapore about his ex-partner, Captain Mike Hatcher.

Australian customs had impounded forty-seven shipping containers of ceramics under their Protection of Movable Cultural Heritage Act (1986), all of which had arrived through Hatcher's company, United Sub Sea Services. Ong had guessed that Hatcher's silence over the last year was not a good sign. Now he knew. He received news of the Australian government bust with mixed feelings. On the one hand, he was glad that Hatcher had been caught and that Australia's hard-line policy on the protection of shipwrecks was being implemented. On the other, he was alarmed at the details hidden in the press reports: Though the containers had been impounded, the Australians were waiting for confirmation from Indonesia that the artifacts had been exported illegally. Ong knew Hatcher too well to ignore this loophole. The other disturbing detail was the mention of forty-seven containers. That was a phenomenal quantity of any ceramic. The news was exactly what Ong had feared most: competition in a niche market. He was already scared that he had brought too much ceramic to the surface. With Hatcher's trove, there could be twice the amount of shipwreck ceramics for sale. To make matters worse, the statements from the Australian police clearly stated that the artifacts were Chinese porcelain. Chinese porcelain was usually of better quality than Vietnamese and had a more established market.

Ong knew that Hatcher was a master salesman. People bought pots from his salvages as much because of the man's buccaneering charisma as for the pieces themselves. How could he compete with that? Some of the auctioneers he had spoken to had suggested that he needed a front man for this project. Ong knew it wasn't the sort of role he could easily fill. Mensun? The thought would have made him wince. This was his wreck, and there was no way Mensun was going to get the credit.

Until now, Ong hadn't known he was in a race, but now it was obvious that timing would be everything. To get the Hoi An cargo to auction ahead of Hatcher, Ong needed to move fast. The first step was to dry the ceramics, then pack them up and move them ashore, where they could be unpacked and categorized in preparation for the sale. Although nearly complete, the process of desalinating the pottery could not be rushed. Every three days the salinity of each tank was tested and the water changed. Ong double-checked the readings himself; a mistake could be disastrous. He had found that even on the nonporous porcelain of the *Geldermalsen* pieces he owned, a light dusting of sodium chloride crystals had appeared after a while. Salt must have worked its way into the uppermost layer of the glaze. It hadn't affected them seriously, for with porcelain the glaze and the fabric of the pot were inseparable and impossible for the salt to penetrate. But these Vietnamese pots were stoneware, and the firing had not had the same unifying effect. The ceramic was ever so slightly porous, which meant that the pots had absorbed salt water. Had they been dried out without first being soaked in fresh water, the salt would remain inside the pots and crystallize. Salt crystals that formed underneath the glaze would render it opaque, or perhaps even crack it. His whole cargo could become worthless.

Though Mensun and other experts had assured him that this level of treatment was excessive, Ong was leaving nothing to chance. He wanted to know optimal temperatures for the desalination. He wanted to be sure that there was nothing in the fresh water that could contaminate or react with the glazes. So, as he patrolled the decks, he was stunned to find Mensun working on a line of artifacts in the bright sun, each of the container lids removed.

"Mensun! What are you doing? The water in these tanks will fill with algae if you let the sun get to them!"

Mensun laughed. "No, no. Ong, look, these pots are high-fired ceramic. There is the risk of salt crystals, but nothing else. Even if any algae did grow in this tank in the short time that I've been working on it, it'd just wash off. You'll see. Relax, Ong, I know what I'm doing. Trust me."

Mensun looked up at Ong from his chair with a forced smile, his eyes dreamy from the painkillers he was taking for his back. With Hatcher returned to the scene, the last thing Ong needed was a cargo of algae-stained dishes. Though furious, all he could do was stare down at the archeologist, his forehead ridged by a pulsing vein.

Part Three

CHAPTER 22

The Task Ahead

IN FEBRUARY 2000, Ong received the news that he'd dreaded. The Indonesian police investigation into Hatcher's salvage of the *Tek Sing* had not been able to prove that the export was illegal. Permits appeared to have been issued, and the Australian police were forced to release all the containers of porcelain, which Hatcher promptly shipped to Europe. There was now no doubt that Ong was going to have to compete with Hatcher at auction.

The ceramics from the Hoi An wreck had also been packed up into shipping containers and moved, but only a few miles inland to two large warehouses in the center of Quang Nam province. With the desalination process complete, the ceramics had been dried and packed into Styrofoam-filled cardboard boxes according to the grid and layer in which they'd been found, and then transported ashore in the containers. When they were unpacked in the warehouses each piece was labeled with a small, numbered sticker on its base. An iconic Chich Choe bird was used as a watermark behind the artifact number, while around its border read the names of the organizations involved: Oxford University MARE, Saga-VISAL, and the National History Museum of Hanoi. The stickers would trace each piece back to its find date and location via a central database and match a certificate of authenticity. Once labeled, the pieces could be freely redistributed around the shelves, according to their form, first of all, and then according to their design.

In a corridor in the main warehouse, between the six-foot-high shelves that ran along its length, Mensun sat at a desk. He looked from the small ceramic box in his left hand to the one in his right, then back again. Both were painted with pavilions two stories high; one of the roofs had a sharper pitch, however. Did they constitute different types of building? Or were they representative of different styles of painting? The artist was not the same. However, if he could find a box lid with a thin-roofed pagoda painted by the same artist who had done the thick-roofed one, he would know that they represented two distinct types of building. He placed the boxes on the table in front of him among a crowd of others gathered between loose mounds of note-covered paper and peered into the shelf beside him. Hundreds of boxes of the same size and shape lined the shelf. A slightly bigger version filled the shelf below, while the three shelves above were taken up with subtle variations on the shape or subsidiary design.

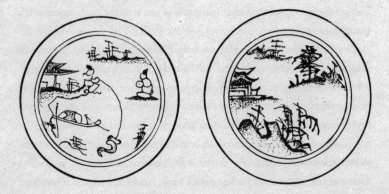

Every few days Mensun's desk moved farther down the corridor, according to the category he was studying. In fact, because he was analyzing several different design types at the same time, he had three or four of these stations spread out around the warehouses, each covered with pottery and piles of paper. He didn't have time to keep tidying them up, and he had given strict instructions that no one else try.

The two adjoining warehouses were filled with quiet activity—the hum of murmured conversations and the clinking of basketfuls of pottery as the last pieces were unpacked from their boxes, labeled, and distributed onto their shelves. Striplights dangled from the ceiling high above, their light watery in comparison with the sunlight that blazed through the doorway at the end of the first warehouse. The silhouette of a guard dominated the entrance, his peaked cap and shouldered rifle casting a large shadow across the dusty floor. He was not guarding against intruders so much as opportunistic laborers. With sixty people working on the cargo in the warehouses, neither Ong nor the Vietnamese government was taking any chances. Everyone leaving the buildings was searched.

The warehouses were no strangers to security. They had been designed to house Quang Nam province's Municipal Rice Reserve. When the rice crops failed, the warehouses became the grain equivalent of Fort Knox. The site had been a Japanese airfield during World War II and was on high ground, beyond the reach of the region's frequent floods. Fences topped with barbed wire surrounded the perimeter while the walls of the three-hundred-foot-long buildings were ten feet thick. None of the security measures could deter the rats that accompanied Mensun during his late-night vigils in the warehouse, however. He could hear their claws scuttling on the dusty concrete floor.

Two years earlier, Mensun had known little about Southeast Asian ceramics; now he was faced with the pressure of writing a comprehensive typology of Vietnamese wares. It was a huge challenge: Scholars of Chinese porcelain often devoted their entire lives to studying one aspect of their subject. But for Mensun, this was the pleasure of his profession. Some archeologists were content merely to record the results of their excavations, then stand back and let others do the analysis. Others brought in specialists for every different aspect. To Mensun, taking on an excavation meant becoming the foremost expert in the relevant field, something he had already achieved with Etruscan shipbuilding and with Tunisian black wares. The analysis, the conclusions, the thinking—to him these were the stimulating parts of his job.

In his book *Vietnamese Ceramics,* John Guy, the Southeast Asian ceramic expert whom Mensun and Ong had consulted before starting the excavation, dates the Chu Dau–style pots to between the late fifteenth and early sixteenth centuries. Mensun had accepted this estimate when he started his journey into Vietnamese pottery, though reluctantly. Since then, he had been comparing various stylistic elements of the Chu Dau pots with pieces from çargoes that were securely dated and begun to think Guy's dates might be wrong. He was becoming convinced that the Vietnamese had been producing this style of ceramic more than half a century earlier, about 1450. Despite all his efforts, however, he had no incontrovertible evidence to prove this. The seabed had not given him enough to establish that one of the foremost experts in the field might be off in his dating. Even Ivo's search through the *Dai Viet,* the chronicles of the Vietnamese kings, looking for mentions of solar eclipses that coincided with large consignments of ceramics, had drawn a blank. He'd searched only the indexes to save time, but Mensun held out little hope that scouring the enormous original volumes would help.

The dating issue aside, the Hoi An cargo had exceeded Mensun's expectations on many fronts. The Vietnamese golden age, hinted at in songs and poems, had indeed truly existed. The excavations at the kiln sites of Chu Dau and other villages had already suggested this, but the pottery around him in the warehouses offered the first showcase of its beauty and variety.

Most of the pieces depicted a simple life, daily labors, and an abiding appreciation for natural beauty. The Vietnamese evidently celebrated the relative freedoms of Buddhism and Taoism, the very faiths that the Chinese authorities had tried to extinguish. The craftsmen who created the Hoi An cargo enjoyed enough liberty to express this in their work. Mensun felt a possessive joy when he walked among the shelves. This cargo was his baby. He knew he had access to it for only a short while, but felt a great responsibility: He was going to be the one to present it to the world. The thought that someone else might publish something first would sometimes wake him in the middle of the night; he would turn on the lights and start working once more. When the auction was done and the pieces were out in the museums and showrooms of the world, it would be open season for other scholars. And that was the problem. If other academics published first and got things wrong, the mistakes would be repeated by others and reverberate throughout the literature. Mensun remembered bitterly the writer who had published an article on the structure of the Etruscan ship at Giglio, erroneously thinking that the keel (perhaps the most important element for diagnosing the nature of a shipwreck) was two-thirds shorter than its actual size because it had still been mostly buried. A similar mistake had occurred on an Elizabethan wreck off Alderney in the Channel Islands, when the rudder—another vital structure—was analyzed without its lower section. As a result, whenever Mensun wrote about either of those ships, both of whose excavations he'd directed, he was forced to include a correction for someone else's mistake, a mistake that worked against his ethos of advancing knowledge. Mensun was determined that the same thing would not happen with the Hoi An cargo. Any mistake so early on in a field of study would propagate rapidly. Only he, the excavation director, had the full perspective of the site, so the responsibility remained with him.

The enormous quantity of ceramics, many of them with shapes and designs never seen before, made the task before him almost superhuman. Somewhere in his brain was the suspicion that it might not be possible to compile the comprehensive typology of the ceramics that this cargo warranted in the time he had left. The typology was a catalog of every single shape of ceramic they'd encountered in the cargo, with thorough descriptions of every category of decoration that adorned those shapes. Each such category was illustrated by several models that had been photographed from every angle and their primary motifs accurately rendered by the corps of draftsmen, and at the base of each section was a long list of artifact numbers detailing exactly which artifacts belonged to this type. Maps charted their distribution across the wreck and allowed patterns of lading to be interpreted. Even a basic version would fill two large volumes. Completing it in less than a year would be Herculean. Mensun had a choice: Make sure he got there first by lightening up on the analysis, or take the time to include all the detail that he wished and just hope that no one published before him.

The more he thought about it, the more Mensun realized that this was too important to rush. His publication would be the standard reference on fifteenth-century Vietnamese ceramics for a long time to come, perhaps for decades or more. Getting it right was crucial. He was the only one with access to the entire cargo, so there was less of a risk that someone might publish behind his back. He decided he would satisfy the Vietnamese government with a basic excavation report. Ong would be granted his export permit and Mensun could continue to work in his own time. As an archeologist, he knew better than anyone the importance of the written record. History would judge him for this final publication, not on the amount of time it had taken him to produce.

Mensun had a more immediate concern, however. Ong had invited the Vietnamese Artifact Evaluation Committee to the warehouse. Made up of museum officials, university department heads, members of Hanoi's Archeological Institute, and the Ministry of Culture, the committee had been assigned by the government to decide which pieces should remain in the museums and which could be sold. Though there was obviously far too much material for one or even one hundred museums to put on display, deciding which pieces were to be sold would be painful. Six out of every ten would leave the country. Ong would be using every bit of his charm to bend things his way. Mensun felt he had

a duty to make sure the Vietnamese weren't persuaded into giving away the best of their country's heritage.

As Ong and Mensun were preparing for the Artifact Evaluation Committee to assess what could leave Vietnam and what must stay, Mike Hatcher was steaming out to the Gaspar Straits to help himself to more of the *Tek Sing*. Though he'd already managed to get forty-seven containers' worth out to Europe, he evidently felt there was a market for more. Either that or he was holding out hope for a cache of gold or some other charismatic treasure that would help him sell the rest of the *Tek Sing*'s cargo. Whatever the case, he'd come equipped for business. Alongside the ever-present *Restless M*, Hatcher had contracted the work barge *Swissco Marine II* and fitted it out with brand-new surface-supply commercial diving equipment, complete with a decompression chamber. With the wreck in less than one hundred feet of water, sixteen divers would rotate shifts on the seabed, decompressing at the end of every shift. Three local vessels had been hired to act as support.

Australian Troy Raven was one of the three divers who went with Hatcher to set out the anchors for the barge and take an early look at the site. He hadn't worked with Hatcher before and was astonished at the scale of the wreck when he reached the seabed. The fifteen bulkheads that divided up the hull were clearly visible, while entire cargo compartments lay stripped and bare but for a few dishes that had become jammed in the waterlogged wood. As in the case of the Hoi An wreck, what remained of the compartments was a good eight feet deep; in those compartments that were still full, the ceramics were stacked on their edge with little else acting as packing, though occasional chests of tea were found, within which the more delicate pieces had been placed.

Work started immediately, with the crews unaware that the Indonesian presidency had changed hands since the previous season. Suharto was no longer at the government helm. Contracts that had been established under the previous regime were no longer worth the paper they were written on. One day, Troy Raven was digging in the spill area to one side of the hull with an airlift. He recalls, "I came upon these little statuettes that were really beautiful. I was trying to be real careful, picking them up real gentle and putting them in the basket." His instincts were right, for the statuettes (chopstick holders in the form of cherubic baby boys with painted blue aprons and finely modeled genitalia) were later appraised by Hatcher's auctioneer as worth between $3,000 and $5,000 each. At the time, however, Hatcher was watching Raven's careful recovery on the monitor on deck.

"What's that cunt doing?" Hatcher shouted at the supervisor, Raven remembers. "Tell him to get working! He's not a bloody archeologist!" The message was passed down to Raven, who looked around him and saw the other divers shoveling porcelain into their baskets.

"No wonder they were doing four bins to my one," Raven later said. "I soon learnt. I was just a baby diver then. I hadn't done much."

Hatcher's team had been working for three weeks when deck crews saw an Indonesian patrol boat cut across the horizon. It had almost disappeared from view when Raven noticed it turn and head back toward the salvage flotilla. As the boat drew close inflatable launches were dropped in the water, and in minutes the Indonesian navy was streaming aboard the vessels, herding the crews onto the foredeck while brandishing AK-47s and shouting. All the crew's rooms were raided for valuable items and cash. Then some of the navy started stripping valuables off the men's bodies. When the crane driver's watch was torn off, he started shouting. The watch had been his father's, and he demanded it back. Suddenly, AKs were cocked and leveled in his direction. Everyone fell silent, knowing that if one shot was fired, all hell would break loose.

The situation gradually cooled off as the crane driver surrendered his claim to the watch. The navy commander demanded to see the boss, but the captured salvage team could not oblige. Just the day before, says Raven, Hatcher had left the site with the tug, heading to the mainland to "sort a few things out." Several of the crew suspected he'd been tipped off that the game was up. The navy ordered the barge to leave her anchors on the seabed and escorted her and *Restless M* to a nearby military port, where they were impounded and the two shipping containers of ceramics that the divers had recovered were confiscated. The crews were placed under guard, detained on board the barge.

Hatcher never returned to Indonesia to try to spring his crews from navy imprisonment, though Raven claims that one of his investors came out to try to help. After a while the navy's guard got more lax. The crew began to be allowed on shore and into town, where they would drink in the local bordellos. There they hatched a plan to escape. They sent all their bags ashore with someone they'd met in town. The next day they were allowed onto land—two months into their detention—they all fled to the airport, where they were reunited with their bags and swiftly departed. The barge *Swissco Marine II* remained in custody for thirteen months, while *Restless M* was eventually returned to Hatcher, having been stripped to the hull. All the sophisticated sonar, magnetometers, fish finders, sub-bottom profilers, GPS, and navigation computers were gone, as well as all the mattresses, cooking utensils, and lightbulbs. Even the taps had been removed.

CHAPTER 23

The Mandarin's Revenge

UNAWARE OF his rival's troubles, Ong welcomed the eight members of the Artifact Evaluation Committee to the warehouse, grinning at them as they stood blinking at the rows and rows of shelves that extended into the gloom. They were among the first outsiders to see the scale of the cargo that had been recovered. Ong wanted them to be impressed: Over the next few weeks they would be selecting which pieces he could take to auction.

The committee soon came to a realization. When the deal between Ong and the government had been struck, they'd estimated that 100,000 artifacts would be raised. Two and a half times that amount was on the shelves of the warehouses, and the 30 percent share allotted to the Vietnamese government for their museums would now amount to more than 75,000 artifacts. Since the financial windfall of the Vung Tau cargo, the government had known that money could be made from the shipwrecks off their coast. However, initially, the Hoi An wreck was considered to be different because the cargo was not Chinese but Vietnamese. Given its rarity and national importance, all of their share of the ceramics were to be retained for the museums. Seventy-five thousand pieces was a different matter—if kept, they would swamp every museum in the country. The committee soon suggested that they should select what they wanted to keep and then sell the remainder at auction, along with Ong's and VISAL's share. The money they made, the committee eagerly explained, would be put into building a new museum wing where a permanent exhibition of the Hoi An cargo could be housed.

Ong tried to look enthusiastic when the idea was put to him, but inside he was in knots. He was already worried that he'd recovered too much, and that his and VISAL's share alone would flood the market. Adding the government's portion would increase the risk substantially. Unfortunately, his export permits had yet to be granted, and Ong guessed these would suddenly become difficult to obtain if he refused the committee's request.

There was one positive aspect: With the government having a stake in the success of the sale, the committee would be more likely to release the significant, high-value pieces for the auction and thus help to attract better buyers. Though the salvage contract specified that every unique piece should be kept by Vietnam, there was latitude in the definition of "unique." A bottle was a bottle, Ong argued, so the museum should keep one bottle as representative of all the bottles. Mensun countered that there were at least twenty different shapes of bottle, some of which carried variations of another twenty designs. The government should keep one of each pattern, he argued, amounting to more than thirty varieties in the case of that category. A subtle battle for influence began, one that would last for the duration of the committee's visit. According to Mensun, who knew the shelves intimately by this stage, the better-quality examples of designs kept disappearing to the back row of their section. He would bring them back to the front in a prominent position so the committee would notice them, only to find the pieces returned to the rear a

few days later. He suspected that Ong, or an
agent working for him, was behind the deception.

The committee's decisions had assumed even
greater importance after Ong's meetings with
the auction house Sotheby's. Their top ceramics specialist, Henry Howard-Sneyd, visited
Vietnam to evaluate the cargo and come
up with a proposal for the sale. Howard-
Sneyd had been overwhelmed at first.
He had admired Vietnamese pots only
as individual treasures in evaluation
rooms; seeing two warehouses full of
them was astonishing. Once he had recovered, he reported to Ong that he
would be delighted to take on such an unusual cargo with such an unusual story. But

there was a catch, he added. Obscure and oversized, the Hoi An cargo
presented a marketing problem. To start with, few people knew about
Vietnamese ceramics. Fewer still had glimpsed the artistry, and not
many of these people collected it. At most there were one hundred
collectors of Vietnamese pottery worldwide. Second, even among the
cognoscenti of Southeast Asian art, Vietnamese ceramics were seen as a
poor cousin to Chinese porcelain. Somehow their skepticism needed to
be transformed into enthusiasm.

Howard-Sneyd's proposed solution was for Mensun and him to give
a series of seminars before launching a two-phase sale. A small selection
of flagship pieces would be put up for auction in a limited collection designed to pique the interest of experts and raise the prices. Once word
had spread, a second, mass-market sale would follow. Here too there was
a snag. Given the large quantity to be sold at this second auction, if the
lots were small they would take weeks to pass through a traditional auction room, yet grouping many pieces into each lot would raise the prices
and deter private buyers. There was another way, however. The Internet

was becoming a viable place to buy and sell. Like other auction houses, Sotheby's was interested in establishing a presence in this booming marketplace and had just founded an online sales department. Ong liked the thought of an online auction to complement the live sale, appreciating that millions of people could be reached through the Internet. There were some issues, of course, involving technology and logistics, but on the whole it seemed a clean and efficient solution to the problem of selling the enormous quantities of ceramics he had recovered. Everything hinged on impressing the influential collectors and museum specialists who attended that first stage, the live auction, and it was the cargo's unique pieces that were most likely to do that. If they were all retained for the Vietnamese museums, Ong would be left with only the more run-of-the-mill pieces, and the whole sales strategy would collapse.

In the end, a compromise was reached on the issue of the definition of unique: Each variation on a shape would be regarded as distinct, but the designs they bore would not. The rule would not be hard and fast, however, and the fate of many pieces would be determined on a case-by-case basis.

CASTING HIS EYE along the rows of shelves in the warehouse, Ong was reminded of the showrooms in Amsterdam where he and Hatcher had displayed the *Geldermalsen* cargo some fifteen years earlier. Indeed, the shelves were familiar for the simple reason that they were of identical design, Ong having commissioned them himself. Now he was about to go to auction again, but this time Hatcher was his competitor, not his partner. It was a fact that made Ong nervous, especially because he had just received some more news on the subject. The German auction house Nagel had announced that they would be auctioning the cargo of the *Tek Sing*. Though Hatcher's second salvage season had been dramatically interrupted, the 350,000 pieces of Chinese blue-and-white porcelain from the previous season were already in Europe.

The race was on. There were some aspects that Ong hoped might play in his favor. Germany had never had a tradition of collecting porcelain, unlike Holland, France, and Britain, and the fact that Hatcher was selling through Nagel was most probably a result of the controversial nature of his operation. Thanks to sustained pressure from ICOMOS (International Council on Monuments and Sites) and UNESCO, auction houses were being forced to consider the origins of what they were selling as never before. A salvage without archeological consideration was one thing; illegal plunder was quite another. Nagel was an unknown on the international auction scene, its sales before that year totaling only $20 million, less than a single painting might fetch at Christie's or Sotheby's. Hatcher claimed that he'd decided against the big players because of the price-fixing scandals that had just rocked the auction world. "They've got their hands tarnished a bit," he said, while a pro-Hatcher news piece in *Business Week* online commented that "tiny Nagel Auctions of Stuttgart, Germany, has pulled off a coup that shows just how far clients' disgruntlement with Christie's and Sotheby's has spread." Hatcher's marketing push had begun. The *Geldermalsen* cargo had come with a dramatic story, and Hatcher's researchers had found another for "the Chinese *Titanic.*" Then Nagel announced that they would be building a replica of the junk in the Stuttgart train station where pieces from the *Tek Sing* would be displayed (to the sound of a specially composed *Tek Sing* theme tune). Traveling exhibitions would also tour major cities in Australia, Asia, Europe, and North America.

The ace in Ong's hand was archeology, but he was beginning to feel that in this game aces were low, not high. Unhindered by the concerns of archeology, Hatcher's salvage must have been many times cheaper. And even though the operation might have been unethical, he was still able to sell the cargo. Without the UNESCO convention on Underwater Cultural Heritage having been ratified, Germany was under no obligation to follow the guidelines that restricted trading in antiquities, and the Indonesian government could do nothing but send a diplomatic letter of appeal to the German authorities.

Again handicapped by the constraints of archeology, Ong did not have the same freedom as Hatcher to apply a name to the Hoi An wreck.

His had been a scientific operation, and thus unsuitable for unproved speculation. No one knew what the original name of the Hoi An ship had been, nor exactly when she had sailed, and there were only clues as to her nationality. While working on the site, everyone had referred to it as the Cu Lao Cham wreck, after the nearby islands, but the name was more obscure and less marketable than "Hoi An." Calling it a "cargo" seemed too closely related to the moniker "Nanking Cargo" that Ong and Hatcher had applied to the *Geldermalsen*. Ong therefore eventually decided that the auction would be of the "Hoi An Hoard." His selling points were the greater historical value and age of the Hoi An Hoard, as well as the fact that it had been excavated with far greater archeological care. He no longer wanted Mensun to carry those flags for him, however. Though they'd brought the cargo up together, Ong had secretly decided that he wanted someone else to endorse it for him.

ONG STALKED the aisles of the warehouse, his expression turning sour. Compared to the *Geldermalsen*'s porcelain, many of these pots seemed slapdash, their lines loose and careless. Picking up a bottle, he turned it in his hands, shaking his head. The painter had used only two lines to draw the bird on the side of the piece. The petal borders to the top and bottom were inconsistent. Chinese potters would never have gotten away with such sloppiness. Their production was much more regular. A whole slew of problems began to sprout in his mind. How could he auction large lots of these if the variations were so great?

Ong had been assured by others, Mensun among them, that their individuality was an asset and that the carelessness was both intentional and positive. He couldn't see it. Putting down the bottle, he walked over to his flagship pieces, the grander examples that had been created with more discipline. So long as he could persuade the Vietnamese committees to include these in the sale, he might have a hope.

Ong reached the shelves where the "specials" were stored. He ran his eye over the ranks of kendis, ewers, and jars, and began feeling more optimistic. It was true some were a little crude. Nonetheless, they did have a degree of charm. Among the pieces he was most enthusiastic about were the four mandarin statues. The experts he had consulted in auction houses and museums had seen or heard of only one or two of these before. Averaging their estimates, Ong reckoned each to be worth about $50,000. Two were chipped, but the breaks were old, therefore inflicted during the sinking rather than by the fishermen's rakes. He picked up one of the better examples and stared at it. Even this piece had been painted sloppily, he reflected bitterly. The eyes didn't match, and the painter hadn't even been able to follow the line of the mouth, which drew upward in one corner. It affected the dignity of the piece.

Turning the statuette in his hand, Ong suddenly froze. Holding it up to the light with both hands, he looked closely at the robes. A yellow stain was creeping up from the base. He licked his thumb and rubbed at it. The stain remained. Fury began to rise within him. One of the most valuable pieces in his cargo was being ruined. Something was wrong with the conservation procedures Mensun had put in place. And if something was wrong with this mandarin, what about the rest of the pieces? He walked quickly down the dimly lit corridors of blue-and-white pottery that now seemed to crowd in on him. In his imagination, yellow stains were creeping through the glaze of every one of them.

"WHERE'S MENSUN?" Ong hissed, suddenly before me in one of the corridors of the warehouse, a mandarin statuette in his hand. He had a strange look in his eye. A string of spittle webbed the corner of his mouth.

"Not quite sure," I replied, cautiously. "Anything I can help with?"

"No. I just want to speak to Mensun."

Eventually we located him in one of the aisles of the second warehouse, squatting among five workers who were going through thousands of small jarlets looking for birds of a certain design.

"Mensun. Look," Ong said, immediately thrusting the mandarin toward him.

Mensun rocked forward from his squat, resting a hand on the ground as he pulled himself to his feet, grimacing as he did so. Bracing one hand at the base of his spine he straightened up, and gradually the creases of pain smoothed off his face.

"What's the matter?" Mensun asked at last, carefully taking the mandarin from Ong's hands.

"Look at the back. Near the base. What's that? *Mmm?*"

"Oh yes. The yellowing." Mensun held it up to the light as Ong had done before him. "Interesting."

"Interesting? Interesting? Mensun, something is going very wrong here. Something is wrong with how we have treated these pots. If this is going to happen to all of them . . . well, we're fucked."

Mensun raised his eyebrows. He brought the pot down and took the reading glasses that he wore around his neck on a piece of string and slid them onto his nose, then reexamined the piece.

"Ong, this isn't caused by us," he said, after a pause. "This could be due to anything. Most probably he was lying near something ferrous on the seabed. The rusting caused a reaction in the glaze. This piece had this stain when it was recovered, I'm sure of it."

"It did not. I have pictures of all these, pictures that I took to have the pieces valued. There was no stain."

"Ong, I think if you go and look you'll find there was a stain before. I remember it now. It may have become slightly more pronounced, but it was always there." Mensun's voice, though still conciliatory, was now edged with anger. His expertise was being questioned.

Wanting to ease the tension, I hurried off and found the original photos. There was indeed a slight yellowing at the base of the statuette. Showing this to Ong did little to lighten his mood.

EVERY PROBLEM served to polarize my two directors even further. Among the last group of artifacts to be unpacked were the boxes. Each box was made up of a lid and a base, while some also had a tray that fitted inside. There were almost fifty thousand artifacts listed as "lid," "base," or "tray" in the records. Therein lay the problem. Though occasionally boxes had been found fitted together on the seabed, the majority (even those that were within the compartments) had been separated from their other sections, sometimes by up to fifteen feet.

Ong wanted to initiate a big matching operation to create complete boxes for the sale. Mensun thought this was a mistake. No evidence suggested that pieces matched with each other had originally been part of a pair, and combining them might mislead those studying the cargo in the future. Ong wanted a complete box to carry a single artifact number that referred to the combination of lid, base, and tray. They would remain as one; having more than one reference number would therefore be confusing and inconvenient for the sale. Again, Mensun strongly disagreed. To keep all the information associated with them together all artifacts must have their own artifact numbers, and bases and lids were separate artifacts. A complete, three-part box would therefore have three numbers associated with it. For Ong, the artifact number doubled as a sales reference. For Mensun, its academic purpose had to remain pure.

THE STICKERS THAT carried the artifact numbers on the base of the pottery were so omnipresent, so often looked at, checked, and double-checked, that they became legible at the merest glance. Their background design no longer registered, and neither Mensun nor I noticed that a terrible mistake had been made.

At the very beginning of the year 2000, MARE had been moved under the umbrella of St. Peter's College, where Mensun held a research fellowship. It was no longer the university body that it had been since its inception. The move was for the most part positive; the Institute of Archaeology that had for so long been trying to rein Mensun in could no longer have any influence on what projects MARE accepted. The negative side was that the unit would have to drop the word "university" from its title, Oxford University MARE, and become simply "Oxford MARE." One day, after discussing the state of university politics with

Mensun, I was checking a dish's artifact number when my eye lingered on the small sticker beneath it and the legend that circled its border: "Saga—VISAL—National History Museum in Hanoi—Oxford University MARE." The bottom of my stomach fell away. The university's name was being used on each and every one of the 250,000 pots in the warehouse. Since the early 1990s, Oxford's lawyers had been getting stricter about the use of the institution's name—especially if it involved commercial purposes such as a sale. Now that it was no longer a university institution, MARE had no excuse for using the university's name on the bottom of the pieces going to auction.

Replacing the stickers presented a logistical nightmare. A new series of stickers was printed that did not feature MARE at all; the rolls filled ten boxes. Finding a single six-figure reference number to match an artifact was time-consuming in the extreme, and had to be repeated for each of the pots that were scattered in random numerical order throughout the two warehouses. The spectre of Hatcher's auction loomed large and Ong threw everything at the problem to get it fixed fast, hiring a new team of workers to help, and managed to complete the replacement and only delay the schedule by a further two months.

IN EARLY MARCH, while Mensun was back in the United Kingdom visiting his family, Ong called me at the warehouse. There had been a change of plan. Though Sotheby's had already spent time examining the cargo and working out a detailed sales strategy, Ong had decided at the last minute to switch to a different auction house.

The new company turned out to be Butterfields, an auctioneer based in San Francisco known mainly for selling arms and armor. Sotheby's was shocked. They had known that Ong would be fishing for deals with their competitors, but had thought it would be with one of the other big, established houses like Christie's, or perhaps Bonhams or Philips. Given that Vietnamese ceramics were all but unknown among collectors, a big name would act as a guarantee. Without it, buyers might question why one of the big names was not selling the cargo. Sotheby's also wondered

at the choice of location. Europe, and particularly London, was the center of the oriental trade ceramics world—the Europeans, with their
colonial past and interests in the Spice Islands, had always had a particular fascination with blue-and-white porcelain. There was no such tradition in the United States.

I was confused, too, until Ong explained matters. Butterfields had recently been bought out by the rapidly expanding Internet auction house
eBay. By far the largest percentage of the cargo was going to be sold online. And if anyone was going to be able to pull off such a huge online
auction—the biggest single antique collection yet to be sold on the Internet—it was eBay. San Francisco was the center of the mushrooming
dot-com boom, and there was more disposable income in the area than
almost anywhere else in the world. Moreover, after Vietnam itself, the
United States had the largest Vietnamese population in the world, and
San Francisco was home to a large proportion of them. The auctioneers
would target them locally, while eBay could reach millions of others
throughout the world via its existing customer base. They wanted to
promote a new prestige auction feature of their site, and had an enormous budget with which to do it.

Sotheby's had suggested running a small, showpiece auction first, to
persuade collectors and experts of the value of Vietnamese pots, and
follow it later with the bulk Internet auction. Butterfields decided to
run the two halves one after the other, in order to maintain marketing
momentum. This made sense to Ong, and he bowed to their opinion.

Though he didn't mention it to me, the clincher for Ong's choice of
Butterfields must have been their terms. The top auction houses
charged far more than Butterfields for hosting the traditional portion of
the sale, and for the online portion eBay was able to offer an unbeatable
deal. It seemed that once again Ong had found a cheaper way. If he
could pull the auction off using Butterfields and eBay, he would end up
a much richer man than if he had sold the cargo through Sotheby's.

At the end of our phone conversation, Ong casually mentioned that
someone would be coming to visit the warehouses. Still trying to take in

his announcement about Sotheby's, I didn't hear the visitor's name properly. We hadn't shown anyone except officials around the warehouses yet. I raised an eyebrow, but of course agreed.

The man arrived that afternoon, as scheduled. He stepped from a jeep with a nervous smile. In a light Australian accent, he explained that he had been working on a land excavation nearby when he had been asked by Ong to pop in and have a look at the ceramics. He politely accepted my offer of a cup of tea, but was quite obviously eager to have a look at the cargo. We walked through the large doorway past the guard, and I heard him suppress a gasp.

"Gosh. Gosh. This is . . . incredible. I can't believe how much . . ." He paused, looking down the length of the warehouse, taking in its depth.

"And this is only one of them. There's another the same through here."

He shook his head in astonishment. We wandered the corridors for over an hour, the man jotting occasional notes on a little pad. When we were done, he handed me his card. I recognized the name, but didn't even think of mentioning it to Mensun.

CHAPTER 24

Academic Attack

THE AUCTION of the Hoi An Hoard was due to start the following week and Dessa Goddard, the head of Southeast Asian arts at Butterfields, threw a party with her boss, Patrick Meade. On the terraces of Goddard's back garden in the hills of North Berkeley, the staff of eBay and Butterfields congratulated each other on their preparations for the upcoming sale. Framed by enormous redwood trees, a constellation of lights from San Francisco and the Golden Gate Bridge stretched behind us.

Everyone was excited. The Hoi An auction would be one of the largest that Butterfields had ever handled. Beautiful catalogs had been produced in record time and new software tools developed for the unprecedented online sale.

The buzz in the air was due not simply to the scale of the auction but to its contents. The pieces going up for sale weren't from the estate of a deceased collector but from a five-hundred-year-old shipwreck raised from the storm-blown, pirate-infested waters of the South China Sea.

Ong and Mensun were the guests of honor, positioned by photographers among various groups of Butterfields and eBay royalty.

We ate outside at large, round tables. I sat next to Jenny Hiltunen. Young, pretty, and ambitious, she had come from an Ivy League university to join the Internet revolution and was in charge of eBay's side of the auction. She gasped and shrieked appropriately as I regaled her with tales of skeletons and storms in the Dragon Sea. What an amazing story, she said. It can't help but be an incredible sale.

Later, however, I overheard her in quiet conversation with another eBay executive.

"Don't sweat it," he was saying. "It's not your fault if it doesn't go right."

"Easy to say but you know how it looks . . . ," she replied.

From then on the flashing smiles, flickering candles, white table-cloths, and stellar optimism of the valley spread beneath us seemed somehow tarnished. After the cargo had left Vietnam, neither Mensun nor I had heard much from Ong at all, and we hadn't been involved in any of the preparations for the auction. One day Magnus Dennis called and said he had seen a small advertisement in the back of the Sunday papers, announcing a small display of Vietnamese ceramics. I went along to the gallery in a Mayfair basement, and found a selection of artifacts from the Hoi An wreck. There had been no other press coverage, nor did the ad even mention the word "shipwreck," and when I visited the show it was all but empty. I was disappointed. I wanted a fanfare for the pottery, a royal parade, not a discreet exhibition. How could an empty Mayfair basement compete with the enormous replica junk that Hatcher's marketing team had built at the Stuttgart train station? Reaching San Francisco, however, my disappointment had evaporated in the excited anticipation.

Goddard, Meade, and Ong stood up one after the other to make speeches after dinner. Each described the long road that had taken them up to this sale. Ong was especially eloquent, understatedly playing the embattled foot soldier who had finally made it home, bloodied but victorious. It had been a hard fight, with troubles and near-misses aplenty, but his belief in the pottery and its worth had kept him going. And now we had triumphed. I almost felt teary.

THE TENSION on the evening of Tuesday, October 10, at Butterfields' San Francisco showrooms was building. The auction was to begin the following morning, but tonight there would be a series of lectures and a question-and-answer session.

The cabinets of the main showroom sparkled. Everywhere I looked, the rustic curves and impressionistic designs of the pottery contrasted beautifully with the hard gilt edges of the display cabinets. Tiny jarlets jostled for space underneath glass counters, while ranks of crab-shaped boxes crowded the shelves alongside line after line of bowls, parrots curling around their rims. I felt proud of the pots I had come to know so well. The stellar pieces stood alone in separate cabinets, lit by individual spotlights. In the auction room itself stood one of the mandarin statuettes, exquisitely illuminated. It was as though an old friend had suddenly become a celebrity, looking perfect but detached from his origins.

Mensun's *Lost Ships* documentary, "White Gold of the Dragon Sea," played on a loop on monitors above, his energetic role-playing adding urgency to the atmosphere in the viewing hall. Potential bidders clustered around the counters and peered past the glass, eagerly assisted by the auction-house staff, who handed pieces over for inspection.

Later I stood by the doors of the auction room where the talks would be given and watched as the audience trickled in. They were an eclectic bunch: A slick, silver-haired suit listened reverently to an academic with an eccentric bow tie and a tweed jacket; a group of Americanized Vietnamese in baseball caps parted for a powerful-looking woman as she led her husband to a particular display case.

Into these colorful waters drifted a raft of black suits. The Vietnamese delegation had arrived. Their seniority was evident not just from their presence, but from their girth. They smiled broadly, the backs of their suit jackets shining as they bowed their heads in repeated greetings, all the while maintaining a spearhead formation.

Ong was playing the perfect host, smiling broadly as he made introductions and welcomed his customers. He looked effortlessly smart in a dark, perfectly tailored suit. Mensun, locked in earnest conversation on the other side of the room, was wearing his standard official outfit: white jeans, leather loafers, and a canary-yellow shirt. A striped tie had been added, as though plucked as an afterthought from a bag of old school clothes.

Mensun had been edgy ever since he'd arrived at the showrooms. He had been asked to contribute a brief article about the wreck, the excavation, and the cargo, but when he saw the catalog his face had darkened. Though the two volumes looked beautiful, their title read *Treasures from the Hoi An Hoard—Important Vietnamese Ceramics from a Late 15th / Early 16th Century Cargo.* These were not, of course, the dates that Mensun had suggested for the pottery. Worse still, behind his introduction to the wreck in the preface to the catalog was another, longer article, discussing the historical context and importance of the cargo. It had been written by John Guy from London's Victoria and Albert Museum, whose name I now recognized with a start. He was the man I'd shown around the warehouses at Ong's request.

I didn't read the articles in question until after the auction. If I had, I might have seen that trouble was brewing. Writing about the nationality and route of the ship, Mensun's language was controlled. There were indications that the vessel might be of Thai origin, he said, and a likely destination could have been any of the Southeast Asian Spice Islands. However, he cautioned that the evidence was, so far, circumstantial and unconfirmed. Guy, however, was bolder in his assertions. The storage jars on deck were Thai, the girl's skull was Thai, the wood of the ship's timbers was (probably) Thai, and other pots were Thai. He concluded that the ship was from Ayutthaya (the Siamese capital at the time) and may have been plying the Ayutthaya–Tonkin trade route. Guy also made some comments on the construction of the vessel, suggesting that it was of nonstandard construction, a hybrid of Chinese and Southeast Asian shipbuilding styles.

Mensun was livid. He had been saving his detailed analysis for his main publication and been deliberately restrained in his article, simply presenting the rudiments of what the excavation had discovered. He also discovered that Guy had written at length on the Hoi An cargo in the Asian art journal *Orientations* just a month before. In his fury at having someone else publish before him, Mensun told me he thought that Guy had been given access to his own work on the cargo to help him come to grips with it. I confessed to Mensun that I'd shown Guy around the warehouses, but he was unconvinced. He said that Guy could never have got enough information in that short visit, and cited similarities in the categorizations that he felt were suspicious. He was convinced that Ong had brought Guy in to spite him, and perhaps even handed over a draft of Mensun's unfinished typology. Whatever the truth, basic archeological courtesies had been ignored, he fumed. "You do not go onto an excavation without the excavation director's permission and then write about it without his knowledge."

Dessa Goddard's assistant tried to defuse the situation by explaining that Butterfields had needed John Guy's input on the cargo, as he was the leading authority on Vietnamese wares, but Mensun was adamant. "It's gross academic sabotage," he announced. "It's totally unacceptable, using someone else's research and publishing with it." He had worked for three years on the Hoi An cargo and put his reputation on the line, and now, he felt, an impostor had stolen his thunder by writing the first article on it. It was the most significant excavation he had yet conducted and it had been made to look as though he—who knew the wreck better than anybody else—wasn't enough of an expert to comment on it. What's more, Mensun snarled, not only was Guy jumping the gun but venturing outside his area of expertise. Alongside Greek vase painting, the structure of wooden vessels was Mensun's most prized specialty.

There were clues, however, that Mensun's teeth had been bared before he had seen Guy's essay. Toward the end of his article, Mensun wrote that the "Hoi An wreck . . . represent[s] a huge reservoir of new, grade-A raw information from a single moment in time. It demonstrates the defective nature of so much of the existing comment . . ." As the main author of most of the existing comment on Vietnamese wares, John Guy had clearly been the target of Mensun's barbed words.

DESSA GODDARD'S voice rose above the hubbub of conversation in the lecture hall, calling on the audience to take their seats and welcoming the first of the evening's speakers. Dr. Pham Quoc Quan, the director of Vietnam's National History Museum in Hanoi, made a short speech. Smiling broadly, he thanked the Excavation Committee, the Evaluation Committee, the Ministry of Planning and Investment, the Ministry of Transport, the Ministry of Finance, the Ministry of Defense, the Ministry of Public Security, the Department for Management of Monuments, the Department of Museology and Conservation, the Border Guard troops of Quang Nam, and the Quang Nam Public Security Forces, all in the incantation of a Communist Party official.

When Dr. Quan's speech was done, Mensun stepped up. He approached the podium with long strides and carefully composed himself, nodding to the projectionist to display the first of his slides. He took the audience through the discovery of the wreck and the recovery of the cargo with his trademark mix of verve and intellect, but it was obvious that he wanted to move on to a different subject.

"There has been much discussion," he said, locking eyes with the audience, "including over the last few days in the showrooms here, about the chronology of this cargo. It is undoubtedly important, for while we know so much about dating Chinese porcelains, there is very little for us to reference this cargo with. In fact, because of its rare nature, it is especially important for those involved in the excavation"—he paused for emphasis—"to help pin it down as precisely and as accurately as possible. This project lays an important foundation for much of the future work that will be done on Vietnamese ceramics."

My eyes found the back of Ong's head several rows in front. I wondered if he had he noticed the change in the timbre of Mensun's voice. Mensun's nostrils were beginning to flare slightly as he took the audience through pictures of the coins, the most recent of which was minted before 1408, and the Chinese pieces found on board that were from somewhere between 1436 and 1464, the rough carbon-14 dating of the wood from the hull that suggested 1449 (give or take fifty years). Soon he was squarely addressing the audience, hands gripping both sides of the podium, the picture that filled the screens behind him forgotten.

Goddard, who had been shifting nervously in her seat, was now on her feet, her slender hands clasped in front of her. She couldn't understand what Mensun was doing. Though the event had been billed as a seminar, she had expected Mensun to give the same talk as he had in Los Angeles a few days before. That, she now realized, was before Mensun had seen the catalog, on the cover of which she had committed to Guy's dating. Mensun was now making her look unprofessional. Any uncertainty over the cargo's date was going to hurt the auction. Collectors liked to know the exact dates of objects. When dealing with Chinese porcelains, a ten-year uncertainty was regarded as a little woolly, and here even their rough guess at "late fifteenth/early sixteenth century" was being called into question.

But Mensun was not stopping there. "Given this evidence, I admit I am at best confused at the chronology given by my colleague in his article in this catalog." The audience were shifting in their chairs while in their midst, Ong sat rigid.

Dessa started speaking, but her microphone was not connected. She shot a fiery glance over at a technician in the corner of the room, and mid-sentence the power kicked in. "Thank you, thank you," she said, a forced smile revealing clenched teeth. "Well, that was very interesting, thank you, Mensun Bound, for your introduction . . . to the amazing work that you and Mr. Ong have managed to accomplish out there. But we must move on; time is short and we have much to get through."

She forced another smile. "Without further ado, I'd like to invite John Guy from London's Victoria and Albert Museum, who will talk in some more detail about Vietnamese ceramics."

His jaw set but eyes looking away from the crowd, Mensun gathered his notes and stalked away from the podium, off the stage, and to the back of the hall. I watched him stand there, pretending to be absorbed in his notes while others around him conspicuously kept their gazes averted. The silence broke into staccato whispers as John Guy approached the stage.

"Hello and good evening and thank you, Dessa," Guy began, blinking behind his glasses. "I'd like to talk about the ceramics of this extraordinary cargo, and where they fit into the general pattern of Southeast Asian art and trade routes . . ."

For the next three-quarters of an hour, he gave a measured talk, outlining the general background of ceramic research in Southeast Asia and the Hoi An Hoard's place within it. When he came to the issue of chronology he simply said it was "complex," neither referring to the embarrassing incident earlier nor pushing home his perspective with details of any dates.

Next on the evening's program of events was a question-and-answer session with both Guy and Mensun. When the time came, they stood on opposite sides of the stage, a palpable tension between them. Goddard fielded questions to them both from the floor, dancing like a boxing referee. Before long, Mensun concluded an answer to one of the audience's questions by bending it onto the topic of dating.

"Let's try to clear things up right now. I'd like to ask one of my own questions of my colleague, if I may." He didn't wait for Goddard's permission, but carried straight on, turning toward Guy. "Nowhere in this cargo do I see any evidence for these pieces having been made any later than around 1480. To have categorized them as late fifteenth to early sixteenth century is misleading and damaging to the study. I would like, right now, to ask my colleague if he has any evidence to back this up."

There was an awkward silence. Mensun was staring directly at Guy, whose shoulders shifted uncomfortably under his tweed jacket. Mensun came from a background in Greek pottery, where years of rigorous thought was backed up by irrefutable archeological evidence. The fixed dates pegged down the periods of relative chronology, and made the field of Greek archeology a bastion of high standards and intellectual discipline. Southeast Asian art history was loose by comparison. Until now, Mensun hadn't realized how strongly he objected to its lack of precision.

Guy blinked as he made notes of Mensun's points. He seemed flustered and confused. Museum experts never confronted one another's expertise in public. Ong had warned him that Mensun had become very possessive of the cargo, but Guy hadn't expected a direct attack. Mensun didn't own the ceramics or have any exclusive academic rights to them, Ong had assured him. He had been hired as an archeologist, on a contract. The cargo was Ong's to have anyone analyze. Besides, Guy thought, this wasn't even Mensun's field. Three years before, when they'd first met, the archeologist had known next to nothing about Southeast Asian ceramics. He, on the other hand, had devoted much of his life to it.

For an agonizing few seconds, Guy showed no sign of rising to the challenge. He eventually answered in a carefully level voice, reiterating some points he had made earlier. He finished by smiling and suggesting that this was perhaps not the place to discuss the issue.

MENSUN LEFT the podium, fuming and still baying for blood. With the question-and-answer session finished, the audience began to make for the tables of drinks that had been set up, a hum of excitement to their voices. Mensun's wife, Joanna, and I headed toward Mensun, but saw him making straight for John Guy. We arrived just as Mensun began his next assault.

"I would just like to know why you have given this such a late dating. Based on what evidence? Do you have anything? Art historical? Archeological? Scientific? You have avoided the question every time it has been raised. And if you have nothing, then admit it!"

Guy's eyelids were flinching under the assault. Mensun was leaning forward over the smaller man, nostrils flaring. Those sipping their drinks around us had started to pull back, their conversations stalled. Joanna and I stepped forward and held Mensun's arms to calm him and allow Guy to back away, which he did.

"But he must be accountable! He has no evidence! He's put forward that dating simply because it brings the ceramics into line with what he's published before. Which, it turns out, is wrong. He must admit it! He's walked in here, supposedly an expert, and got it wrong. Worse than got it wrong, he's done it with a motive. It's vital that this is corrected. This will be how people remember this wreck, and he's got it wrong! If he is a true academic he should be prepared to argue his case. And you saw, Jo, Frank—you saw. He's just walked away. Twice. He's got no evidence for it!"

CHAPTER 25

Final Reckoning

THE NEXT MORNING, the showroom was teeming with activity. Mensun's documentary still played on monitors over the heads of the crowds, while among them auction-house staff smiled energetically, selling catalogs, registering bidders, and ushering people into the hall. Television cameramen and press circled. At the back, two operators sat at computer terminals to relay all online bidding to the auctioneer. A technician hovered. It was the first time that this kind of technology had been used for such an important auction. Opposite the computers was another row of staff, all smiling or talking earnestly into telephone receivers as they coaxed long-distance bidders into the buying mood.

By ten o'clock all the seats were taken. Newcomers were forced to stand at the back of the hall. I could feel the excitement building in myself; looking over at Ong, I could tell he felt the same. Patrick Meade, the director of Butterfields, stepped up as the first auctioneer. Exuding the smooth charm integral to the job, the large Irishman announced the first lot. It was a wide dish with an ornate, barbed rim, but badly stained and with only traces of the original enameled design. I held my breath as the bidding began, but it turned out I was in no danger of suffocating. The bidding stopped at just over half the lower estimate of $5,000.

The next dish, featuring a beautiful, gleaming blue-and-white depiction of a lion chasing a brocade ball through the clouds, fetched $32,500—well above the higher estimate. From then on, the auction seemed to pick up pace. When we broke for lunch I was exhilarated. It was all going well, though I could not find Ong to congratulate him.

The afternoon session continued to coax high prices from buyers. The dragon ewers, featured in a line drawing on the cover of both catalogs, came under the hammer. Though I couldn't help feeling resentful at the pots I'd come to know so well being sold in this way, now that the auction was running I wanted everything to fetch as much as possible. The bidding for the dragons was frantic, the atmosphere tense. Subtle glances passed between the various leading players, representing the world's museums: Phoenix, Santa Barbara, New South Wales, Denver, Los Angeles, Seattle, Singapore, and the British Museum, among others. Though Butterfields estimated their hammer prices would be between $30,000 and $50,000, the three dragons sold for $79,500, $63,000, and $57,500, the last two being bought by the British Museum and the Gallery of New South Wales.

THE NEXT DAY the auction hall was not quite as full. The momentum had somehow slowed. The lots were passing faster, and by the end of the day the gleeful timbre of the auctioneer's voice had become apprehensive. Some lots weren't attracting many bids at all. A mysterious phone bidder was grabbing the pieces that no one else wanted—Mensun suspected this was the auction house "bidding against the chandelier," protecting itself against selling too low.

Mensun regretted that the reception for the cargo wasn't electric. It would have been gratifying to see the public react in the same way that he had when he had first seen that grinning mandarin. On the other hand if the auction had turned into another *Geldermalsen* the attendant publicity would have been enormous and while he had done a good job, given the circumstances, and extracted a huge body of information from the cargo, his typology was not quite ready for publication. Until it was, his detractors would have a field day. Less honorably, it also didn't pain Mensun to see Ong get stung a bit. Though Ong was putting a brave, professional face on things, Mensun knew his profit-obsessed partner would be getting worried by now.

On the third day, Mensun himself began to worry. The hall was less than half full for most of the day. Even the Vietnamese delegation had stopped smiling and was looking anxiously at the empty seats around them. Not having a runaway success was one thing; a total disaster was quite another.

For the first time in two years, Mensun was concerned about funding. Ong had always been there to provide. Decompression diving didn't work? We'll use saturation. Need more space? We'll bring in another barge. More warehouse staff? Done. But if the auction crashed, Mensun's research funds would dry up overnight. With every unanswered call from the auctioneer, the lavish, full-color, hardcover, two-volume publication he had been planning on became less certain. This was serious. The published typology was the only way of presenting our findings and showing his critics that he had fulfilled his responsibility. The books were dry and clinical, but they had to be. They were presenting a huge volume of information in an organized, intelligible, and comprehensive way and he was justifiably proud of the work. For the first time Mensun realized that the prices fetched in the auction were vitally important for him as well as for Ong. He began to will the prices upward.

As the last few lots were announced on the third day, Ong blinked slowly. His dream had become a nightmare. He had allowed himself to feel triumphant while speaking at Dessa Goddard's party, and had basked in the adulation of the Butterfields staff. Now those same staff averted their eyes in embarrassment. Only 40 percent of the lots had sold. The pyramid of future projects he had built up in his mind crumbled.

Not only had he lost his personal stake, he had lost the trust of his investors. Without that, he had no way of making money.

Ong had known that shipwreck salvage was risky business. That was partly what had attracted him to it. But somehow he had lost sight of the stakes. He had gambled hard against the weather by using budget equipment. He had lost that round, and then doubled the bet when he brought in a saturation system. Using old, run-down equipment had been another risk, but there he'd come out on top—the operational phase of the job had been completed for a fraction of the expense that Dilip had first budgeted. At every stage, Ong had found a way to cut costs dramatically. That's how he had managed to get this far. Now, at this last hurdle, the cost-cutting had caught up with him. He had spent almost $14 million recovering the Hoi An cargo and bringing it to auction. The sale had brought in just $2,960,000.

Ong left the auction hall in a daze. Where had he gone wrong? Had he made a mistake by choosing the lower rates charged by Butterfields rather than paying more for a more established auction house like Christie's or Sotheby's? Was the United States the wrong market? Had he been misled about the potential interest in Vietnamese pottery?

There were some factors outside Ong's control that had taken their toll. EBay had bought Butterfields in 1998 for $210 million, a price that was way over the odds. That had not worried the rich technology company overmuch. Perhaps the most successful dot-com (and during 1999, the only profitable one in existence), eBay had wanted the Hoi An Hoard as a showpiece auction to promote their acquisition of Butterfields. But when the NASDAQ index began to fall in 2000, eBay was forced to tighten its belt substantially, and Butterfields' marketing budget was slashed. A culture clash between the MBAs at eBay and the Ph.D.s at Butterfields worsened the situation, ensuring a lack of synergy between the two that further stretched the funds.

Ong was as yet unaware of any of this. As he passed the open door of the showroom, video screens were still flickering inside, showing a scene with Mensun pretending to guide one of the recovery crates to a safe landing on the deck of the tug. Ong's lip curled. He found it bitterly ironic that the man he had employed as an archeologist in order to ensure a good story for the cargo had, in his eyes, only ensured a good story for himself. Ong had ended up funding a film promoting Mensun Bound, not the Hoi An cargo. Of the publications that were supposed to be out in time for the auction, he had seen nothing. Then, to cap it all, Mensun had gone one worse. In front of the assembled buyers, he had actually questioned the authority of the one person who could still add credibility to the cargo.

The element of shipwreck recovery Ong had thought would give him the edge had taken it off: the archeology. It had extended the expensive excavation phase and added all the drawing, photographing, numbering, and renumbering of the post-excavation work.

Salt was rubbed into the wound when Mike Hatcher's porcelain cargo from the *Tek Sing* went under the hammer a month later. The press reports screamed that the eight-day sale was a huge success, bringing in a total of DM22.4 million ($14.5 million), with all 16,100 lots selling. More than five hundred Internet bidders (from Germany, the United Kingdom, the United States, Taiwan, Hong Kong, Singapore, Japan, and even Uzbekistan) had resulted in 13 percent of the lots selling online. To Ong, the message was clear: The public didn't want sunken treasure to be squeaky-clean and scientifically recovered; they wanted swashbuckling, controversial adventure.

WITH DISAPPOINTING PRICES set during the live auction of the Hoi An Hoard, the later online sale through eBay followed suit. Seen on the sterile medium of a computer screen, the work of the Vietnamese potters stood no chance. Both professional and amateur collectors were used to touching the pieces they were about to buy; it was how they fell in love. The small, thumbnail pictures on the Web site conveyed none of the charm of the pieces. Moreover, the disclaimers were worrying, warning that pots might be cracked, chipped, or have a more faded glaze than in the pictures they were viewing. The millions of eBay users Ong had envisaged swamping the auction site did not materialize, and, just as in the live auction, a few knowledgeable collectors mopped up the desirable pieces at knockdown prices. Perhaps a few years later more enthusiasts might have trusted the Internet, but such was not the case in 2000. In his own way, Ong had been taken in by the dot-com bubble.

With the flagship pieces now gone from the collection, cherry-picked by the savvy institutions and individuals who weren't put off by his choice of auctioneer, only the less attractive pieces remained. Further tainted by the failed sale, they became even less desirable and, unsold, the ceramics were left to languish in rented storage space in a warehouse outside San Francisco, their cost increasing slowly but surely. Moving them to a cheaper location outside the States would mean yet another round of import and export taxes. Ong was at a loss for what to do.

Mensun and I continued to work hard to bring the typology to completion over the next few months, but once the time came to make the first payment to the company that had laid out the twelve hundred pages for printing, Ong stopped answering our calls. He no longer had any incentive to pay for the publication; as far as he was concerned, the project was bankrupt. Mensun and I were left open-mouthed, but helpless. We had a two-volume, full-color publication ready to be printed at a cost of almost $200,000, substantial debts, and no financial backer.

Because of MARE's ongoing political problems with the Institute of Archaeology at Oxford, Mensun knew more than most the importance of publishing his work. As he wrote in a guide to underwater excavations for the Royal Geographic Society's Expedition Advisory Centre in 1990, "in broad terms field archeology is all about new information and it is the archeologist's duty to disseminate this at a scholarly level through publication. To not do so is the height of irresponsibility . . . Excavation is destruction."

Nonetheless, Mensun remained unwilling to compromise on what he believed the material deserved, and refused to be rushed into producing a cheaper version. With no deadline pressing, he became more and more convinced of the importance of using the information to create a publication that did the wreck justice. Simply publishing was not enough in itself—after all, Hatcher had produced a book (written by his researcher, Nigel Pickford) on the *Tek Sing*, but that didn't make the salvage any more ethical. Publishing papers in academic journals could never cover the breadth of material—in Mensun's eyes, such reports were simply a way of deferring a final comment. It wasn't until more than twenty years after the excavation of the *Mary Rose* that the final report was printed, he reasoned. He could only hope that over time, what he considered the truth about the chronology of the Hoi An wreck would come out and that the field would not suffer too long under any misconceptions caused by Guy's high dating of the cargo.

MENSUN AND I weren't the only ones to suffer from the auction's failure. According to Dilip, Ong left bad debts in both Singapore and Vietnam before relocating to Perth, and it is claimed by some dealers that Butterfields still holds a substantial collection of the ceramics as collateral against unpaid bills.

EBay eventually sold Butterfields, which was then acquired by the British auction house Bonhams in 2002. Having joined forces, Butterfields & Bonhams soon became the third-largest auction house in the world. As for eBay and Internet auctioneering, the failed online sale of the Hoi An pots was not the ill omen it seemed to be. EBay's online auctions are now the second-biggest earner on the Internet after the search engine Google, operating in seventeen countries and making $3 billion in 2004 alone.

IN EARLY 2001 in Adelaide, Australia, a special meeting of shareholders in Mike Hatcher's Ocean Salvage Corporation was called. The company directors had decided to come clean about the real financial results of the *Tek Sing* operation. The managing director, Stephen McNamara, stood up and told shareholders that despite all the publicity to the contrary, the company's salvage of the *Tek Sing* had barely covered costs. The auction was "not successful," it was reported in the *Adelaide Advertiser;*

189,000 pieces had not sold (presumably, as in the Hoi An sale, these were simply bought back via bids to the chandelier). McNamara explained that the operation had cost approximately $20 million and though the company was hoping to make back between $25 and $45 million, they'd ended up with only $17 million (up from Nagel's figure of $14.5 million, presumably due to follow-on sales). The company chairman, Kevin Burrowes, then stood up to admit that after the cost of locating, recovering, and selling the *Tek Sing*'s cargo, the company was worth "diddlysquat," and its irate shareholders were owed nothing.

Most treasure-hunting companies operate using the same formula. Find a shipwreck with a stirring story and a manifest that mentions treasure, and then sell the dream to shareholders. The treasure-hunters may or may not believe the dream; what is certain is that all the shareholders' money will be spent on the search, whether it is successful or not. Inflated salaries and friendly backhanders from boatyards who charge the earth for refits are all par for the course. Even when the project is a disaster according to the accountants, the treasure-hunter rarely loses out. Compared to the $14 million that Ong apparently spent on the Hoi An salvage, it's hard to believe that the *Tek Sing*—at half the depth and without the added cost of archeological control—cost $6 million more. Indeed, even the figures for Hoi An seemed suspicious to some. When talking the finances over with Dilip Tan, I mentioned the stated barge-running costs of $50,000 per day and the total of $14 million. Both figures, according to the operations manager, looked at least twice as high as he estimated they should have been. Ong's company being private, this is not something that can be proved; but if Dilip's estimates are correct, and if Ong really did charge his investors $14 million, then Ong's metamorphosis from accountant to treasure-hunter was truly complete.

THE INDONESIAN GOVERNMENT eventually managed to prove that the *Tek Sing*'s porcelain had indeed been illegally exported, but by then it was too late: The sale had come and gone. Indonesia's only reprieve came in September 2001 when the Australian Federal Police discovered seven more shipping containers of porcelain from the *Tek Sing* in Adelaide, and with the wrongdoing now proved, confiscated them and sent them back to Indonesia.

CERAMICS FROM the Hoi An cargo soon began appearing in mail-order gift catalogs alongside porcelain from the *Tek Sing*, as Ong and Hatcher saw their respective treasures offered together in bundled deals as though they were discounted seconds. I can only assume that Ong, private about his business affairs at the best of times, has now gone back to more traditional investments. I do know that his memories of our adventure in Vietnam remain fond, however. I met with him in London some years after I'd last seen him in San Francisco. I wanted to talk to him about his plans for the ceramics that were gathering dust in San Francisco, hoping that somehow we could come up with a plan that would at least part-finance the unpublished typology. As we walked through the streets of Bayswater looking for a place to eat, we both noticed a Vietnamese restaurant on the other side of the road. I veered toward it, he veered away, and we collided on the sidewalk. We ended up eating pizza.

Ong later put *Tropical 388* up for sale, stripped of the aging saturation system that had been taken by the dive contractors as part of their deal. The barge was eventually bought by none other than Mike Hatcher, who had by no means given up his profession. The treasure-hunter's name appears in the transcripts of a Singaporean courtroom session in 2005, where the owners of a vessel called *Seeker 1* alleged that he had been responsible for their ship being impounded by the Indonesian authorities after implication in an illegal salvage operation off their coast. Though concrete news of him is hard to come by, Hatcher was last rumored to be searching for a saturation system, presumably to place on his new barge.

The rise in oil prices in the years following the Hoi An excavation meant that such systems were in high demand, and so were the crews needed to run them. Hartmut Winterberg took a job with Shell, advising them on their diving operations. The other members of the dive crew employed by the dive contractors moved on to new contracts with barely a backward glance. Chad Robinson is now a sought-after saturation diver, able to be selective about which jobs he takes—he still works with his father whenever he can. Jack Ng, though he had disliked his first taste of saturation in Vietnam and left the system early with Roland Dashu, decided to give it another try and is now a fully fledged saturation diver. "I had to cut my hair," he explained to me over the phone, having just pulled his new black Porsche 911 off the highway in Singapore. "There were a lot of complaints from the guys in sat. You can't have long hair. You can't imagine the hassle. Now just short, normal Chinaman."

Garth Sykes still takes his pick of the best supervising jobs that come up around the world, while Tony Turner has become a businessman, running hyperbaric medical centers in Brisbane and Sydney, and a research institute as well. Nigel Kerr eventually realized that although diving on shipwrecks might be fun, it would never pay him the salaries that were being offered for his military experience. He returned to Algeria, where he leads security teams that protect foreign installations and personnel. His teammate Hector George took the same turn. When I last spoke to George, he told me he was working as the personal bodyguard to the British ambassador in Iraq.

Dilip went back to working offshore, taking oil, gas, and fiber-optic cable installation contracts in China, Taiwan, Hong Kong, the Philippines, Malaysia, and Brunei. Though he still supervises projects, he chooses to spend six months of the year working in saturation. His affair with the world of historic salvage was over. He'd been driven to working with Ong because he was fed up with the predictability and consistency of normal commercial operations. Now he reveled in those same qualities.

Having left his job as a management consultant to work for Ong, Ivo returned home to the Czech Republic to pursue his interest in the history of language. As a member of the prestigious Prague Linguistic Circle and the Czech Linguistic Society, he continued teaching and writing with the traditional fervor of a Czech intellectual, producing books and papers while also finding time to adopt the Vietnamese daughter of an artist he had befriended during the Hoi An project, whose education he now oversees.

The Vietnamese government was disappointed at the auction's financial failure but proudly filled the nation's museums with the Hoi An wreck's ceramics, establishing permanent exhibitions in both the Hanoi and Ho Chi Minh City branches of the National Museum of Vietnamese History as well as in the Vietnamese Fine Arts Museum. Quang Nam province boasted a display, while a whole new ceramic exhibition house was built at the Hai Duong Museum in the area that had produced the wares.

THE ISSUE OF imported antiquities remains a burning one. Museums have long been under pressure to return relics—such as the Elgin marbles in the British Museum, or the Ishtar Gate in Berlin's Pergamon Museum—but have so far resisted claims for repatriation to the countries of origin, in these cases Greece and Iraq. They are under increasing scrutiny over the provenance of new purchases, however, scrutiny that led to the curator of the John Paul Getty Museum in Los Angeles being arrested in December 2005, accused of buying stolen antiquities. The pressure from marine archeologists is particularly intense, with the result that many American museums refuse to display artifacts recovered from shipwrecks, no matter what the provenance, for fear of encouraging treasure-hunters. Sadly, however, the UNESCO treaty on underwater artifacts has not come into force, and there are always other avenues for sale. While developed countries can afford to take the ethical high ground and while major auction houses and museums have reputations to uphold, there's always a newcomer eager to make a name for itself.

With the market still open, the treasure hunt continues. A few operators choose to work with sensitivity. In 1998 a German mechanical engineer named Walterfang Tilman heard about a shipwreck site from his Indonesian in-laws. He quietly recovered the exceptionally rare ninth-century Tang-dynasty wreck in a controlled fashion, then held out against all offers of splitting the cargo, choosing instead to keep it in one place as the centerpiece for a new maritime museum on Singapore's Sentosa Island resort. In most of Southeast Asia, however, it is still open season on the seabed. Though the allure of porcelain is somewhat tarnished by the disappointing results of both the Hoi An and the *Tek Sing* auctions, it's common knowledge that auction houses are still on the lookout for the next *Geldermalsen*.

And the ships are out there, their demises recorded and their remains waiting to be found. Among the estimated three million wrecks lying on the seabed around the world, shining like a beacon to eager young treasure-hunters is the Portuguese ship *Flor do Mar*, which sank in the Straits of Malacca in 1512 after Admiral Alfonso de Albuquerque had sacked the port of Malacca. Her cargo, according to Albuquerque, was "the richest treasure on earth." The loot, pillaged from the sultan and the city's rich merchants, included two bronze lion sculptures that had stood guard over the palace (a gift from the Emperor of China) and the ruler's golden throne. Sixty tons of other gold treasure was accompanied by more than two hundred chests of diamonds, emeralds, rubies, sapphires, and other precious stones. Tentatively valued at $19 billion, the *Flor do Mar* is the most valuable treasure wreck on record. Contrary to the reports that appear like clockwork in the press (each a thinly veiled attempt to raise funds for a search), she still lies beneath thick mud and beyond the reach of technology, at least for now.

DURING THE HOI AN project, Mensun and I had often spoken about what we were going to do next, and even put together a detailed proposal for a South African company that wanted to salvage a porcelain cargo off Cape Aguilas on the southern tip of the continent. The waters at the junction of the Indian and Atlantic oceans are notoriously dangerous, terrorized by storms above and by great white sharks below. The multimillion-dollar saturation operation we proposed (though never taken further) gives an indication of our thinking at the time. Things changed after the Hoi An auction. Mensun sat me down one day and explained that he no longer wanted to expand MARE. He was scared of turning into an administrator, tied up with paperwork and legalities. He would rather be working in the field, in touch with the projects and the wrecks that he was investigating. My own ambitious hopes with MARE were crumbling.

Not knowing what else to do, I joined Mensun on one more wreck in Mozambique, but a spectre of commercial interest lurked behind the funding for the project. Uncertainty over the eventual fate of the cargo and the information we were collecting robbed me of the passion that had made recording wrecks a pleasure. I somehow felt dirty, as though I was legitimizing what the treasure-hunters were doing. For a while I felt let down by Mensun, for so long my mentor, believing that despite all his passion and commitment he'd sold out. Then I realized there was a difference between us. Mensun was independent. He didn't have to please treasure-hunters by cooperating with their whims, and he didn't have to wait for the approval of academic committees or bow to scholarly cliques that he felt were misguided. By contrast shipwrecks for me weren't just about information and knowledge; I was trying to make my living from them. In essence I was just another treasure-hunter.

Mensun heard my troubled thoughts with good heart. "Listen," he told me. "There must be a commercial dimension on certain wrecks because otherwise they will never be done and will be destroyed instead. The money is just not there in archeology, learned societies, or third-world governments to do this work. Besides, 95 percent of archeologists are not field people—though they like to think they are—and less than one percent are prepared to work in far-flung impoverished countries."

I nodded, but with a heavy head. The rationale didn't change my situation, and the Mozambique project was my last. Disillusioned with shipwreck archeology and confused by the compromises it involved, I decided to step ashore to grapple with the landlocked life that I had managed to evade for so long.

EPILOGUE

The Gibbon & the Sun

LONG AFTER THE DUST had settled on the Hoi An project and its aftermath, I went to Mensun's house outside Oxford on a summer evening. Mensun had a surprise visitor staying with him. There in the front room was Ivo Vasiljev, hunched over a Nintendo game doing his best to bend his thumb to the unfamiliar game pad under the tuition of Mensun's youngest son. He'd come to England to participate in a medical project—researchers were scanning the brains of linguistic masters to try to discern the roots of their extraordinary abilities—and he thought he'd pay us a visit.

After many warm greetings, we retired with Mensun to a picnic table outside. Ivo leant forward in his chair, hands clasped in front of him eagerly.

"I have what might be exciting news! Well, *I* am excited!" He was smiling broadly and his eyes were bright. "Do you remember the dish that I showed you both on *Abex,* the one that carried a painting of a tree with circles beneath?"

We both nodded. The piece featured a one-off design that had caught all of our imaginations with its depiction of an eclipse, for it offered a slim prospect of a date for the cargo. Mensun was looking serious, as though he was wracking his brain.

"Ivo," he said. "Didn't you already check the *Dai Viet*?" The chronicles of the Vietnamese king were the most likely place that both a solar eclipse and the arrival of a foreign vessel would be recorded.

Ivo admitted that his first investigation of the *Dai Viet* had turned up nothing promising, but that he had only gone through the index volume at that stage. Knowing that the original tomes were enormously lengthy and that the indexer had not been on the lookout for the same things as he, Ivo had subsequently gone back to the originals and painstakingly gone through the entire fifteenth and sixteenth centuries looking for significant coincidences.

"I found I think five or six eclipses of the sun during that period," Ivo said. "They were always recorded, for the Vietnamese were very superstitious. An eclipse was a big event, an indication that a big disaster or a big opportunity was about to come to the kingdom.

"Now, three of these eclipses coincided, within a few years, with notes on the arrival of foreign ships. This is perhaps strange. The bays must have been full of foreign ships. Indeed some poets say as much. So I think those ships that were mentioned were large, important, or otherwise significant."

Ivo began to explain. One of these vessels, from Sumatra, had visited in 1467 carrying some Chinese passengers. This created a diplomatic problem for the Vietnamese king and he sent these Chinese émigrés back to the Ming. Another vessel from Siam was recorded to have arrived the same year. This ship was carrying an envoy from Thailand who was bringing gifts to the Vietnamese king. There was another diplomatic problem, however, and the envoy was dismissed, his gifts rejected. Thirty years earlier, in 1437, there was a third coincidence. This was the one that had excited him.

"In 1436 there was a very famous eclipse. It is famous because of the story of an important Daoist priest who blackmailed the king with his knowledge of astronomy," Ivo said. "In 1434 the high priest, who was also an astronomer and so knew that a total eclipse was coming soon, approached the king and warned him that there was danger ahead. A great gibbon was going to eat the sun, he claimed. It would of course be disastrous if an ape ate the sun. The priest told the king that if he was paid lots of money he could prevent the gibbon from gobbling." Ivo chuckled at his wordplay, and went on. The king, knowing the implications to his kingdom of losing the sun, consented. The priest then made a show of organizing a great gibbon hunt, and hundreds were rounded up from the forest. Lo and behold, that year no gibbon ate the sun; the eclipse never came. The king was overjoyed, but the following year when the priest once again issued a warning, the ruler was dubious. He decided not to pay the priest to round up gibbons. But as the priest knew, this was the year of the eclipse and the sun was indeed consumed—although only for two minutes. Any doubts over the authenticity of the priest's warnings were quashed, and he lived out his days as a trusted advisor. The king never did realize he had been cheated by the priest's knowledge of astronomy, Ivo explained. That was an analysis made by later historians.

In 1437, the year after the eclipse, the chronicle recorded a Thai ship arriving, then, three months later, the visit of a delegation. The mention of this deputation in the Dai Viet had been accidentally omitted from the index volume, and it was here that Ivo's dedication had paid off. The delegation had been accompanied by an envoy from Thailand who was evidently good at his job, for the chronicle recorded that the king had rewarded the emissary lavishly and halved the export taxes that he normally inflicted.

"Out of every hundred items," Ivo explained, "the king would usually take twenty as tax. But here he cuts this to ten. And he also sends gifts to the king of their country and to his consort; many rolls of painted silk and thirty sets of porcelain to the king, fifteen to his consort. The king would only be so generous if the delegation had come to negotiate something really big. And this cargo was really big. Also there is the fact that there is no record of such a big shipment ever arriving in Thailand, meaning it must have disappeared on the way."

There was a problem, however. "Why the gap between the entries?" I asked. "Surely the deal should be struck first, before the ship arrives? Or at least with the ship? Otherwise they might run into diplomatic trouble and have wasted the journey."

"Ah. Well, this is the other part of my, um, hypothesis." Though bashful, Ivo was almost trembling with enthusiasm. "My theory is that this ship originally went to China, where her owners were expecting to pick up a large shipment of porcelain that they had already taken orders for. This was a common way for the Chinese to do their business—otherwise the trade was too much of a gamble. So, they turned up, but when they arrived they got a shock. The Ming government had banned exports, again. The Ming ban of 1436. Nothing was allowed to leave the country; in fact the ship was not even allowed past the river mouth.

"Desperate, her owners turn south. They head to Van Dong, the major port at the mouth of the Red River delta. They've heard rumors that there is a ceramics industry there. And there is. So the ship arrives, then has to wait for the order to be fulfilled before it can continue its journey, also dependent on the winds."

It was a plausible explanation, but I now had another objection. There were only a few months between the entries in the chronicles, and it seemed unlikely that all of the pots we had seen on the wreck could have been made in three months, let alone taking into account those that the fishermen had taken.

"Nonsense! This is possible. There were perhaps three-quarters of a million pots on the boat, but there were over two hundred villages in the Chu Dau area alone. It was a huge industry."

Mensun looked dubious, and pointed out that they would have been able to fire the pieces only in the dry season.

"And where are the other pots they made?" I said. "If they were making a million pots every three months then Southeast Asia, if not the world, should be full of them!"

"You are not a businessman!" Ivo countered. "When there is demand the supply will appear!"

"What, in the space of a month?"

"The industry was already there, it was just manufacturing other types of ceramic, like tiles for temples. When there's suddenly a huge order, they won't turn it down. The king himself has bent over backward to sweeten the pot for the trader. They will not let him down."

"So this was a one-off production?"

"Exactly. They had never done anything of this scale before, and would not do afterward. Of course they would manage to sell smaller numbers of similar ceramics afterward, but never anything like what they'd done here."

Mensun grinned. He'd always thought that this cargo was extraordinary. Ivo's evidence was not of a type that he could use in an academic debate—there were too many suppositions—but it warmed him nonetheless. Although John Guy's initial dating of the cargo to the late fifteenth—early sixteenth century was still being repeated, as Mensun had known it would be, Guy had now revised his thinking on the subject and estimated the Hoi An wares to have been made between 1470 and 1480. The Vietnamese were bolder in their assertions, according to Ivo, who had recently attended a ceramics conference there. They now believed that the wreck dated from the 1430s or 1440s, putting the Hoi An trove squarely in the middle of Vietnam's golden age.

Some years later, I went to an exhibition of the Hoi An Hoard at the British Museum in London. As I was leaving, out of the corner of my eye I caught a glint of light from a display cabinet that I'd overlooked. I turned and saw before me the kneeling mandarin, polished and resplendent, grinning at me. I grinned back. The years had distilled my memories of the project and now the statuette's amused gaze reminded me of the legends from its era. One told of a boy who became a dragon, benevolent but for the wrath he reserved for the greedy. Another inspired monuments to the Sea Dragon whose fearsome jaws would strike only if the pearl they gripped was stolen. For a long time I stared at the mandarin, pondering Ong, Mensun, and the potential of divine intervention from some ancient god, wondering if the superstitions of the Vietnamese might have some currency after all.

Eventually I walked away, out of the museum and into the crisp sunlight. Standing there on the sidewalk I felt the years of resentment about the turbulent end to our journey through the Dragon Sea evaporate. Now I understood that the lesson of the Hoi An cargo was not about the folly of greed or ambition but was instead a broader revelation of human nature. Ong and Mensun had both believed that theirs was the single true path, and I'd always assumed that only one of them could be correct. Now I knew that there was no absolute truth but instead an infinity of possible directions and that in choosing a course only your own heart could be trusted for guidance. Looking across the street at the flow of civilization, life suddenly seemed lawless and uncontrollable, as open as the ocean itself.

ACKNOWLEDGMENTS

My life changed with all that happened in the Dragon Sea, though I didn't realize it at the time. The people involved in the excavation of the Hoi An wreck created this story, and it is they that I must thank first. I am particularly indebted to those who were kind enough to speak to me in detail about it afterward: Hector George, John Guy, Jenny Hiltunen, Henry Howard-Sneyd, Mike Hughes, Nigel Kerr, Amanda Miller, Jack Ng, Troy Raven, Chad and Robbie Robinson, Jon Street, Garth Sykes, Dilip Tan, Tony Turner, Steve Webster, Hartmut Winterberg, Woody Woodhouse, and Matthew Wortman. Special thanks to Ong Soo Hin, who was generous enough to talk to me despite the difficult nature of the subject, and to Ivo Vasiljev for enduring unending follow-up interviews. Mensun Bound first invited me into the extraordinary underworld of shipwrecks and exploring it with such a passionate enthusiast has been an experience for which I am extremely grateful. I know that my writing this book has at times been a painful process but I hope,

Mensun, that you feel it was worth it in the end. Many thanks also to Joanna Yellowlees-Bound, for her constant patience and support.

Those who generously helped me understand new subjects were Michael Asher, Roger Box, Tom Cockerell, Diarmaid Douglas-Hamilton, Colin Martin, Robert McPherson, Ian Oxley at the Oxford Archaeological Unit, Dr. John Ross at the National Hyperbaric Centre in Aberdeen, Joe Ruskin (veteran of the DB-29 disaster), Colin Sheaf, Anne Smith, Charles Thomas, Martin Woodward, Reg Vallintine, and of course my father, Maurice Pope, whose unending humor and achievements in his physical and intellectual life never fail to astound.

I regret not being able to revisit Vietnam to interview the Vietnamese members of the team, a fact that has doubtless harmed the text. Unfortunately, given my inability to speak Vietnamese, interviews over the phone were impossible. However, I'm greatly indebted to Nguyen Viet Cuong for his help via e-mail. A tiny fraction of the extraordinary work of the draftsmen from the National Museum in Hanoi and the Institute of Archaeology is presented in these pages. For allowing me to reproduce their beautiful renditions of what their ancestors had painted on the ceramics so long ago, my thanks go to Nguyen Thuong Hy, Nguyen Quoc Huu, Nguyen Viet Cuong, Nguyen Tuan Lam, Nguyen Duc Minh, Ha Nguyen Diem, Nguyen Son Ka, Vien Khao Cohoc, and the late Trinh Viet Cu. My own rough sketches are limited to those that illustrate technical aspects of the operation and general impressions. Photographs from the Hoi An excavation can be seen at www.dragon-sea.com.

The crews of *Abex, Tropical 388, OL Star, Ena Supply 3,* and *Tropical Star* were all welcoming and helped explain the things about barge life that I didn't understand—in particular Nguyen Huu Tai, Johnny Leong, Mahmud, and Dyad. The staff of Hoi An's Thuy Duong II hotel put up with rowdy crews descending on them before and after every season, as did long-suffering Mr. Triet of Hoi An's notorious Triet's Bar. A special nod to Magnus Dennis, a vital ally in those times and ever since. He saw the importance of the "real" story even while the excavation was going on, and raised the man overboard alarm when I disappeared between the hulls of the barges one morning.

Wringing a readable story from what happened in the Dragon Sea was a journey in itself. My eldest brother, Hugh, was solely responsible for convincing me that there was land on the other side of what seemed an uncrossable ocean, and it was he who pushed me to embark in the first place. For help pulling up anchor I must thank Camilla Hornby, Gail Pirkis, and Diana Finch for their early support. However, my trade wind has been my wonderful agent, Claire Paterson, whose brains, belief, cunning, and laughter have made the otherwise agonizing process a pleasure. In the United States, the guru of literary agents, Tina Bennett, showed rash impulsiveness in pushing my manuscript from the start, an instinct for which I'll be forever grateful. Harcourt has been my spice Islands, where Timothy Bent has been full of good humor, faith, and perseverance while expertly molding the story after first terrifying me with a behind-the-scenes tour of New York bookstores. David Hough's copyediting was done with an eagle eye, and Susan Amster steered me past treacherous reefs with charm and skill. After three years at sea, meeting Rowland White at Michael Joseph (Penguin) was like finally sighting English soil; his advice (and Carly Cook's) was as crucial as an architect's to a builder.

I cannot begin to thank all the friends who have helped me along the way. I'm especially indebted to Kate Weinberg for incisive readings, advice, and a crucial introduction, and also to Neil Barrett, Nicholas Evans, Sasha Filskow, Ben Freedman, Will Lacey, Tim Moore, Brian and Prunella Power, Lavinia Thomas, Tilly Ware, Sam Woods, Martin Woodward, Rupert Wyatt, and my brother Hugh and his wife, Jessica. My gratitude once again to my parents and to all my other siblings— Thomas, Patrick, Helen, and Quentin. You were vital to my mental stability, and I love you all.

Thanks to my new parents-in-law, Iain and Oria Douglas-Hamilton, for their encouragement, patience, and unending generosity while I ate them out of house and home. Finally, thank you to Saba, my wife, lover, partner, best friend, and sternest reader, for advice on the importance of poetry, for encouragement, and for explaining to me and others what the book was about when I'd lost sight myself.

Despite the input of all these wonderful people, each mistake is of course entirely my own.

A NOTE ON SOURCES

Most of the events recounted in this story I experienced firsthand, and their details were recorded in my journals and letters at the time. Other aspects of what happened in the Dragon Sea came to light only during an extensive series of interviews I conducted while researching this book. These same discussions also yielded most of the information that makes up the narrative before I joined the project. At times the testimonies of my subjects conflicted. In these instances I took the majority view or, if the jury was hung, the view that tallied best with my own recollection. The substance of all dialogue in the text is as reported in my journal or by my interviewees, but given that I had no intention of writing this book at the time and had no tape recorder running, the exact words are in some cases my own.

The intense monitoring that the divers were under for the duration of their saturation was primarily for their safety, but it also meant that their lives for those fifty-eight days were under constant scrutiny. Though privy to all that they said and did, those of us on the surface were not able to read their minds. Where I have entered their internal world, I have done so through their own recall during interviews or through talking to other saturation divers who have been in similar situations. Commercial divers work in a tight-knit community where confidence in each other's abilities and experience can be a matter of life and death. For this reason, the identities of some of the divers have been changed.

In researching the background for the story I consulted far too many books, papers, and articles to list here. However, for an overview of the accomplishments and scope of archeology in the sea, the British Museum's *Encyclopedia of Underwater and Maritime Archaeology* is a comprehensive beginning. The references they cite will guide interested readers to more information on most of the subjects and wrecks I have touched on. The early works on underwater exploration, such as Cousteau's *Silent World* (1953), Diolé's *The Undersea Adventure* (1954), and Bass's *Archaeology Under Water* (1966), all capture the excitement of discovering a new world on our own planet, while later works such as Marx's *History of Underwater Exploration* (1978), Earle and Giddings's *Exploring the Deep Frontier* (1980), and Throckmorton's *Diving for Treasure* (1977) are all useful summaries. The best starting point for more information on the conflict between treasure-hunters and archeologists is the Web, which can point you to vehement defenses of both ends of the spectrum, together with arguments for every intermediate shade. Shipwreck salvage stories abound, but my favorites are David Scott's *Seventy Fathoms Deep* (1932), Keith Jessop's *Goldfinder* (1998), and Gary Kinder's *Ship of Gold in the Deep Blue Sea* (1998).

Ivo Vasiljev was a human crystal ball who allowed me to browse confidently through the history of Vietnam and China, his knowledge seemingly infinite. Stevenson and Guy's *Vietnamese Ceramics: A Separate Tradition* (1997) was an essential introduction to Vietnamese pottery. Mensun Bound kindly allowed me a sneak preview of his excellent general history and background to the Hoi An wreck, a book in the final stages of production that will accompany the typology mentioned in the text.

INDEX

He just wanted a decent book to read ...

Not too much to ask, is it? It was in 1935 when Allen Lane, Managing Director of Bodley Head Publishers, stood on a platform at Exeter railway station looking for something good to read on his journey back to London. His choice was limited to popular magazines and poor-quality paperbacks – the same choice faced every day by the vast majority of readers, few of whom could afford hardbacks. Lane's disappointment and subsequent anger at the range of books generally available led him to found a company – and change the world.

'We believed in the existence in this country of a vast reading public for intelligent books at a low price, and staked everything on it'
Sir Allen Lane, 1902–1970, founder of Penguin Books

The quality paperback had arrived – and not just in bookshops. Lane was adamant that his Penguins should appear in chain stores and tobacconists, and should cost no more than a packet of cigarettes.

Reading habits (and cigarette prices) have changed since 1935, but Penguin still believes in publishing the best books for everybody to enjoy. We still believe that good design costs no more than bad design, and we still believe that quality books published passionately and responsibly make the world a better place.

So wherever you see the little bird – whether it's on a piece of prize-winning literary fiction or a celebrity autobiography, political tour de force or historical masterpiece, a serial-killer thriller, reference book, world classic or a piece of pure escapism – you can bet that it represents the very best that the genre has to offer.

Whatever you like to read – trust Penguin.